HOST-PLANT SELECTION BY PHYTOPHAGOUS INSECTS

Contemporary Topics in Entomology

Series Editors

Thomas A. Miller
Department of Entomology
University of California at Riverside
Riverside, CA 92521

Helmut S. van Emden
Department of Horticulture
University of Reading
Earley Gate, Whiteknights
Reading RG6 2AT
United Kingdom

Previously published

Plant Kairomones in Insect Ecology and Management
Robert L. Metcalf and Esther R. Metcalf

HOST-PLANT SELECTION BY PHYTOPHAGOUS INSECTS

E.A. Bernays
and
R.F. Chapman

Contemporary Topics in Entomology 2

CHAPMAN & HALL
I(T)P An International Thomson Publishing Company

New York • Albany • Bonn • Boston • Cincinnati • Detroit • London • Madrid • Melbourne •
Mexico City • Pacific Grove • Paris • San Francisco • Singapore • Tokyo • Toronto • Washington

For more information contact:

Chapman & Hall
115 Fifth Avenue
New York, NY 10003

Chapman & Hall
2-6 Boundary Row
London SE1 8HN
England

Thomas Nelson Australia
102 Dodds Street
South Melbourne, 3205
Victoria, Australia

Chapman & Hall GmbH
Postfach 100 263
D-69442 Weinheim
Germany

Nelson Canada
1120 Birchmount Road
Scarborough, Ontario
Canada M1K 5G4

International Thomson Publishing Asia
221 Henderson Road #05-10
Henderson Building
Singapore 0315

International Thomson Editores
Campos Eliseos 385, Piso 7
Col. Polanco
11560 Mexico D.F.
Mexico

International Thomson Publishing - Japan
Hirakawacho-cho Kyowa Building, 3F
1-2-1 Hirakawacho-cho
Chiyoda-ku, 102 Tokyo
Japan

2 3 4 5 6 7 8 9 XXX 01 00 99 98 97 96

Library of Congress Cataloging-in-Publication Data

Bernays, E. A. (Elizabeth A.)
     Host-plant selection by phytophagous insects / by E. A. Bernays and R. F. Chapman.
          P.   Cm.   --(Contemporary topies in entomology; 2)
     Includes bibliographical references (p.   ) and indexes.
     ISBN 0-412-03111-6 (hb).--ISBN 0-412-03131-0 (pb)
     1. Insect-plant relationships. 2. Phytophagous insects--Host plants. 3. Phytophagous insects--Behavior. I. Chapman, R.F. (Reginald Frederick) II. Title. III. Series.
     QL496.B46   1994
     595.7053--dc20
                                                                                  93-25190
                                                                                  CIP

Visit Chapman & Hall on the Internet http://www.chaphall.com/chaphall.html

To order this or any other Chapman & Hall book, please contact **International Thomson Publishing, 7625 Empire Drive, Florence, KY 41042.** Phone (606) 525-6600 or 1-800-842-3636. Fax: (606) 525-7778. E-mail: order@chaphall.com.

For a complete listing of Chapman & Hall titles, send your request to **Chapman & Hall, Dept. BC, 115 Fifth Avenue, New York, NY 10003.**

Contents

2.4.2.3 Flavonoids, quinones and tannins 40
2.4.2.4 Variability in levels of phenolics 42
2.4.3 Terpenoids 44
2.4.3.1 Mono-, sesqui- and diterpenoids 44
2.4.3.2 Triterpenoids 46
2.4.3.3 Variation in levels of terpenoids 48
2.4.4 Organic acids, lipids and acetylenes 50
2.4.5 Sulfur-containing compounds 51
2.5 Conclusions 54

3. Sensory systems 61
3.1 Introduction 61
3.2 The sense of taste 63
3.2.1 Distribution and abundance of contact chemoreceptors. 66
3.2.2 Variability of response 70
3.2.3 Sensory coding 73
3.3 The sense of smell 76
3.3.1 Distribution and abundance of olfactory receptors 81
3.3.2 Integration of olfactory inputs 83
3.4 The sense of touch 84
3.4.1 Distribution of touch receptors 85
3.5 The sense of sight 86
3.6 Conclusions 90

4. Behavior: the process of host-plant selection 95
4.1 Host finding 97
4.1.1 Odor-induced attraction 97
4.1.2 Visual attraction 107
4.1.2.1 Shape and size 107
4.1.2.2 Color 109
4.1.3 Conclusions 115
4.2 Acceptance of the plant 117
4.2.1 Physical properties of the plant 118
4.2.2 Leaf odor 121
up to & including 4.2.3 The plant surface 122
4.2.4 The internal constituents of the leaf 128
4.2.4.1 Phagostimulants 128
4.2.4.2 Pharmacophagy 137
4.2.4.3 Deterrents 138
4.2.5 Making the unpalatable palatable 147
4.2.6 Conclusions 148
4.3 What happens in the field? 150

5. Behavior: the impact of ecology and physiology 166
5.1 The hosts and the habitat 166
5.1.1 Host plants 167
5.1.2 Other plants 176

Preface

For more than 20 years insect/plant relations have been a focus for studies in ecology and evolution. The importance of insects as crop pests, and the great potential of insects for the biological control of weeds, have provided further impetus for work in this area. All this attention has resulted in books on various aspects of the topic, and reviews and research papers are abundant. So why write another book?

It seems to us that, in the midst of all this activity, behavior has been neglected. We do not mean to suggest that there have not been admirable papers on behavior. The fact that we can write this book attests to that. But we feel that, too often, behavior is relegated to a back seat. In comparison to the major ecological and evolutionary questions, it may seem trivial. Yet the whole process of host-plant selection and host-plant specificity amongst insects depends on behavior, and selection for behavioral differences must be a prime factor in the evolution of host-plant specificity.

In writing this book, we hope to draw attention to this central role of behavior and, hopefully, encourage a few students to attack some of the very difficult questions that remain unanswered.

We have deliberately avoided citing references in the text because we feel that the flow of language is very important, and a plethora of references would detract from the flow. Our object is to encourage the reader to read on! We hope that our method of presenting the references will make it easy to delve deeper into the topics and examples in the text.

We wish to pay tribute to three elder statesmen in the field who, largely unwittingly, have influenced our approach, and who, sometimes, more directly, have affected the thinking of a whole generation of students. They are the late Vincent Dethier, whose contributions to insect feeding behavior and chemosensory physiology are seminal; Tibor Jermy, whose recognition of the roles of deterrent chemicals and learning by phytophagous insects have been milestones in this field; and the late John Kennedy, one of the first and very few to have looked at insect behavior in the field and whose rigorous approach to the study of behavior has been salutory to us all.

Many other friends and colleagues have contributed to the book in more direct ways. In particular, we wish to thank the members of Liz's lab who, at an early stage in the writing, generously gave their time to read and criticize the whole manuscript. Their inputs were especially valuable to us because they saw the book through the eyes of its intended readers, from new graduates to postdocs. They are: Kathy Burgess, Jennifer Ciaccio, Jon Diehl, Heather Geitzenauer, John Glendinning, Nancy Nelson, Greg Sword and Martha Weiss.

Louis Schoonhoven has read the whole manuscript and made a number of useful suggestions. Others who have read and commented on parts of the book are: Judie Bronstein, Nancy Moran and Lynne Oland. Greg Sword cheerfully produced all the line drawings and never tired of our demands for something slightly different. Betty Estesen prepared the species index. Phil Evans produced the illustrations of chemical formulae. Terry Villelas typed and retyped the tables. We are grateful to all of them.

Elizabeth Bernays
Reg Chapman
Tucson, Arizona

March 6, 1993

Acknowledgments

The following figures and tables are reproduced, in whole or in part, with the permission of the publishers and/or journals.

Figs. 1.1, 3.9, 3.15, 3.19, 4.24, 4.25, 4.27a, 4.33, 5.11, 6.3, 6.4b, 6.8, 8.1. Table 2.2. By courtesy of Academic Press Ltd.

Fig. 2.2a is reproduced, with permission, from the Annual Review of Ecology and Systematics Vol. 11, © 1980 by Annual Reviews Inc.

Fig. 2.3a. By courtesy of *Oikos*.

Figs. 2.3a, 2.4a. From "Seasonal changes in oak leaf tannins and nutrients as a cause of spring feeding by winter moth caterpillars" by P. Feeny, *Ecology*, 1970. *51*, 565–581. Copyright © 1970 by the Ecological Society of America. Reprinted by permission.

Figs. 2.3a, 6.1, 7.2, 7.9. By courtesy of PUDOC—Centre for Agricultural Publishing and Documentation.

Figs. 2.3b, 2.4a, 4.12, 4.13, 4.16, 4.27, 4.29, 5.2, 5.4, 5.14, 5.22, 6.4a, 6.5, 8.2a. Tables 5.1, 6.2, 7.3. By courtesy of Blackwell Scientific Publications Ltd.

Fig. 2.4a. By courtesy of the Society for Chemical Industry.

Figs. 2.11a, 6.6. By courtesy of Akadémiai Kiadó.

Fig. 2.11b. Reprinted from Biochemical Systematics and Ecology, vol. 5, Cooper Driver et al., Seasonal variation in secondary plant compounds in relation to the palatability of *Pteridium aquilinum*. pp. 177–183, Copyright (1977), with kind permission from Pergamon Press Ltd, Headington Hill Hall, Oxford OX3 0BW, UK.

Fig. 2.12. Reprinted with permission from *Insect Plant Interactions*, vol. 4. Ed. E.A. Bernays, 1992. Copyright CRC Press, Inc. Boca Raton, FL.

Figs. 2.13, 2.27, 2.32, 4.14, 4.26. By courtesy of Plenum Press.

Fig. 2.19. Reprinted with permission from *Insect Plant Interactions*, vol. 1. Ed. E.A. Bernays, 1992. Copyright CRC Press, Inc. Boca Raton, FL.

Figs. 2.21, 5.9. From Phytochemical Induction by Herbivores, eds. D.W. Tallamy and M.J. Raupp. Copyright © 1991 John Wiley & Sons, Inc. Reprinted by permission of John Wiley & Sons, Inc.

Fig. 2.22. By courtesy of the *Journal of Heredity*. Copyright 1989 by the American Genetic Association.

Fig. 2.26. By courtesy of Hodder & Stoughton.

Figs. 2.20, 3.4, 3.12, 4.8, 4.28, 5.6, 6.2, 7.11, 7.12, 8.3. Tables 2.8, 4.11, 5.3. By courtesy of Springer-Verlag, New York, Inc.

Figs. 3.5a, 3.8, 3.11a, 3.13, 3.15a, 3.16, 3.17, 4.5, 4.6, 4.17, 4.23, 5.18, 5.20, 6.7, 7.6, 7.7. Tables 4.3, 4.8, 4.10. Table 6.1. By courtesy of Kluwer Academic Publications.

Figs. 3.5b, 3.10a. Reprinted from Journal of Insect Physiology, vol. 36, W.M. Blaney and M.S.J.Simmonds. A behavioural and electrophysiological study of the role of tarsal chemoreceptors in feeding by adults of *Spodoptera, Heliothis virescens* and *Helicoverpa armigera*. pp. 743–756, Copyright (1990), with kind permission from Pergamon Press Ltd, Headington Hill Hall, Oxford 0X3 0BW, UK.

Fig. 3.7. Reprinted from the Internationl Journal of Insect Morphology and Embryology, vol. 18, R.F. Chapman J. Fraser. The chemosensory system of the monophagous grasshopper, *Bootettix argentatus* Bruner (Orthoptera: Acrididae). pp. 111–118, Copyright (1989), with kind permission from Pergamon Press Ltd, Headington Hill Hall, Oxford 0X3 0BW, UK.

Fig. 3.10b, 4.3, 4.22. By Courtesy of the Company of Biologists Ltd.

Fig. 3.21. By Courtesy of the National Research Council of Canada.

Fig. 4.4. Reprinted from Journal of Insect Physiology, vol. 23, K.N. Saxena and P. Khattar. Orientation of *Papilio demoleus* larvae in relation to the size, distance and combination pattern of visual stimuli. pp. 1421–1428, Copyright (1977), with kind permission from Pergamon Press Ltd, Headington Hill Hall, Oxford 0X3 0BW, UK.

Fig. 4.7, 4.10. By Courtesy of the Entomological Society of America.

Figs. 4.6, 4.9. Reprinted from *Science*. Copyright 1983 by the AAAS.

Fig. 4.15. By courtesy of the Centre National de la Recherche Scientifique.

Fig. 5.3. By courtesy of the *Journal of Ecology*.

Fig. 5.7. By courtesy of E.J. Brill.

Fig. 5.17. By courtesy of CSIRO Editorial Services.

Figs. 7.1, 7.10, 8.8. Table 7.1. By courtesy of the University of Chicago Press.

Figs. 7.3, 7.4, 7.8, 8.12. By courtesy of *Evolution*.

Fig. 7.13. Reprinted with permission from Plant Resistance to Insects, ed. P.A. Hedin. Copyright 1983 American Chemical Society.

Fig. 8.2b. By courtesy of Sigma Xi, The Scientific Research Society.

Fig. 8.4. From "Competition between exotic species: scale insects on hemlock" by M.S. McClure, *Ecology*, 1980. *61*, 1391–1401. Copyright © 1980 by the Ecological Society of America. Reprinted by permission.

Fig. 8.5a, 8.6. By courtesy of Chapman & Hall, London.

Fig. 8.5b. By courtesy of the Royal Entomological Society.

Fig. 8.7. Reprinted with permission from *Insect Plant Interactions*, vol. 3. Ed. E.A. Bernays, 1992. Copyright CRC Press, Inc. Boca Raton, FL.

Fig. 8.11. From "Saturniid and sphingid caterpillars: two ways to eat leaves" by E.A. Bernays and D. Janzen, *Ecology,* 1988. *69*, 1153–1160. Copyright © 1988 by the Ecological Society of America. Reprinted by permission.By courtesy of the Ecological Society of America.

Table 5.2. Reprinted from Journal of Insect Physiology, vol. 34, Honda et al. Fungal volatiles as oviposition attractants for the yellow peach moth *Conogethes punctiferalis* (Guenée)(Lepidoptera: Pyralidae). pp. 205–212, Copyright (1988), with kind permission from Pergamon Press Ltd, Headington Hill Hall, Oxford OX3 0BW, UK.

Introduction

It is common knowledge that plant-feeding insect species make up over one quarter of all macroscopic organisms, and that the green plants upon which they feed make up another quarter. Every extant green plant has insect herbivores. They chew the tissues or suck the juices of all plant parts, in spite of the immense variation in nutrient level, nutrient balance and the many physical and chemical factors that might be expected to provide barriers to attack. Insect herbivores have evolved infinite numbers of paths for exploiting plants, for overcoming their deficiences as food, and for dealing with their many protective devices.

As a consequence of their diversity and abundance, insects are a major link, perhaps *the* major link, between the primary producers, green plants, and a multitude of animals at higher trophic levels. The abundance of insect parasitoids and predators is a direct consequence of the abundance of phytophagous insects, and many vertebrates, from fish to mammals, depend on insects for their livelihood.

Phytophagous insects may also have had a part to play in the evolution of plants, by selecting for diverse chemical and physical defenses. Many ecologists believe that the current diversity of both plants and insects is, in part, a result of their coevolution. Others believe that the adaption and diversity of the insects has tended to follow that of the plants, there being less effect of insects on the evolution of plants.

These evolutionary questions have provided a principal area of focus for many studies of insect/plant relationships. A second area of study centers around the fact that some species are so important in human affairs. Many feed only on a limited range of plants and, with the development of large areas of crops, some of these specialists have become major pests, like the brown plant hopper of rice and the Colorado potato beetle. Some others have become pests by virtue of their lack of specialization; the desert locust and the gypsy moth are good examples. Sometimes phytophagous insects can be used to the practical advantage of humanity. This is the case with some specialist insects that have been used in the biological control of weeds. One of the classical examples is the control of prickly pear in Australia by caterpillars of *Cactoblastis cactorum,* imported from South America.

Behavior tends to have been neglected, perhaps because it may seem trivial beside these important evolutionary and economic questions. Yet, behavior is central to both areas. Behavior must be the first thing that changes when the process of adaptation to a new plant begins. Largely because of the widespread occurrence of chemical deterrents in non-host plants, an insect that encounters a non-host will, in general, not remain for long. Before any physiological or morphological changes that adapt it to the new host can occur, it must overcome this deterrence. Consequently, behavioral changes must be the first to be selected for when an insect invades a new host. Thus behavior, and especially variation in behavior, is central to our understanding of the major evolutionary questions of insect/host plant relationships.

Behavior, too, is central to our understanding of varietal resistance in agricultural crops. Many plant resistance mechanisms are known to be behavioral, although the resistance has commonly been developed without a knowledge of the behavior. Many more could surely be developed with a greater knowledge. Nowhere is this more apparent than in the embryonic field of genetic engineering. Plant geneticists are ready to engineer changes in plants to make them resistant to insects. But what should they engineer? We can rarely answer this question because our knowledge of insect behavior in relation to the plants is so inadequate. It is the details of behavior that are important here, just as they are in trying to understand the natural process of evolution.

So in this book we tell the story of the behavior of insects in their interactions with plants. The two central chapters, Chapters 4 and 5, contain the core of the behavior. We describe the principal mechanisms that an insect uses in locating and identifying its host, and then show how, in reality, these mechanisms must be seen in the context of the environment. Leading up to these chapters are three others providing the background necessary for an understanding of the insect/plant system. Chapter 1 briefly describes the degree of host-plant specificity found amongst phytophagous insects. Chapter 2 tells us about the chemistry of the plants and how it varies. Chemistry is emphasized because it is the key to so many interactions. Chapter 3 gives an outline of the sensory systems involved in the insect's perception of plants. It is framed especially for the nonentomologist to try to give an idea of the capabilities and limitations of the systems.

Chapters 6 and 7 discuss how variation in behavior may arise. The idea that learning may have a significant role in modifying an insect's response to its host is relatively new, but the field is expanding fast. So far, our knowledge of variation in the behavior of individual insects is slight. Traditionally, it has been usual to present data as means and deviations. As a result, information on individuals has been lost. Only recently has the importance of individual variation in the selection of adaptive characters been fully appreciated.

Finally, in Chapter 8, we return to the beginning. Now, we examine the factors that may have contributed to the host-plant ranges that we see today.

1

Patterns of Host-Plant Use

1.1 Host-plant range

A continuous spectrum exists between insect species that will only feed on one plant species and others that feed on a very wide range of plants in many different families. It is usual to separate the insects into categories depending on their host-plant ranges, but it is important to recognize that no clear boundaries separate these groups and different authors use them in different ways. The categories commonly recognized are: monophagous, oligophagous and polyphagous.

Strictly, monophagous means feeding on only one species of plant, but the term is usually extended to include species feeding on plants within a single genus. Oligophagous is used to refer to insects feeding on a number of plants, usually in different genera within one plant family. Sometimes, however, an insect is associated with a small number of plant species from different families. Finally, polyphagous refers to insects feeding on a relatively large number of plants from different families. Alternative terms occasionally found in the literature are stenophagous, referring to insects with a restricted host-plant range, and euryphagous, for insects with a broad host-plant range. We shall not use these terms.

Some authors use monophagy, oligophagy and polyphagy strictly in relation to the numbers of plant species attacked without reference to their taxonomic relationships. However, an insect feeding only on 10 plants from the same family is almost certainly using some characteristic of the plants that they have in common in determining their acceptability, while an insect feeding on 10 plants from different families probably employs a number of different cues. It is more informative to use the terms to infer some uniformity of behavioral response rather than strictly quantitatively, and we refer to such examples as oligophagous and polyphagous, respectively.

Monophagous species occur in all the major groups of herbivorous insects. For example, amongst the grasshoppers (Orthoptera), *Bootettix argentatus* is monophagous on creosote bush, *Larrea tridentata*, which extends through thousands of square miles of desert in the southwest United States and Mexico. This grasshopper is never found on other plants and it has been shown experimentally that it will not eat anything else. It is strictly monophagous.

Many Lepidoptera (butterflies and moths) are also monophagous, amongst them the heliconiine butterflies. These are spectacular insects restricted to the New World tropics. Their caterpillars feed on passion flower vines (family Passifloraceae), and various species of *Passiflora* are hosts for members of the genus *Heliconius*. Some of the butterflies only feed on one species of *Passiflora*, while others feed on several. *Heliconius melpomene*, for example, is specific to *P. oerstedii*, even though in the laboratory it can survive on other *Passiflora* species; *H. cydno*, on the other hand, feeds on at least five species of *Passiflora*. This example illustrates the difficulty of categorizing feeding patterns. For practical reasons we regard both species as monophagous, although obviously one has a wider host-plant range than the other.

Examples of monophagous species amongst the sucking bugs (Hemiptera) are the spotted alfalfa aphid, *Therioaphis maculata*, which, as its common name indicates, only feeds on alfalfa, *Medicago sativa*, and the brown planthopper, *Nilaparvata lugens*, which feeds on rice throughout south and southeast Asia. *Chrysolina quadrigemina* is an example of a monophagous beetle (Coleoptera). It has been used in the biological control of Klamath weed, *Hypericum perforatum*, in California. This practical use has only been possible because the beetle is restricted to feeding on species of *Hypericum*. Amongst the two-winged flies, Diptera, *Dacus oleae* is an example of a monophagous species. Its common name is the olive fruit fly and the larvae only occur in olives in the Mediterranean region. A final example is the spruce sawfly, *Diprion hercyniae*, whose only hosts are spruce trees in north-temperate regions of Europe and America. It is a representative of the order Hymenoptera which includes bees and wasps and many parasitic insects in addition to a small number of phytophagous species (see Fig. 1.2 below). Other examples of monophagous species will be found in the text.

A good example of an oligophagous insect species is the Colorado potato beetle, *Leptinotarsa decemlineata*. It is recorded as feeding on 14 plants, all in the potato family, Solanaceae, and mostly in the genus *Solanum*. The related beetle, *Leptinotarsa rubiginosa*, is even more restricted. It also feeds on solanaceous plants, but only on two species of *Physalis* and two of *Solanum*. However, some oligophagous species may feed on much larger numbers of plants if the group on which they specialize contains large numbers of plant species with characteristics in common. For example, many grass-feeding grasshoppers, like the migratory locust, *Locusta migratoria*, will eat many, perhaps hundreds, of different grasses because all of these grasses possess features in common that are used by the insect in selecting its food. Sometimes an insect may eat plants from different families because they possess some chemicals in common. For example, the larva of the cabbage butterfly, *Pieris rapae*, mainly eats plant species in the family Brassicaceae, but it will also eat nasturtium, family Tropaeolaceae. These plants are characterized by the possession of glucosinolates. We still call these species oligophagous because the term tells us that the host-plant range is restricted to plants with some particular characteristic.

The term "disjunct oligophagy" is sometimes used to refer to the feeding pattern of an insect that feeds on a very small number of plants from different families even though we do not understand the connection between these different plants. The grasshopper, *Ligurotettix coquilletti*, from the southwestern United States and Mexico is a good example of a species exhibiting disjunct oligophagy. For most populations, the creosote bush, *Larrea*, family Zygophyllaceae, is the host, but populations are also known to occur on *Atriplex*, family Chenopodiaceae, and on *Lycium*, Solanaceae.

We use the term "polyphagous" to mean insects that feed on plants from more than one family which, at least on the basis of our present knowledge, have no common feature which governs selection by the insect. Very often, this involves a large number of plants from several different families, but there is no boundary between disjunct oligophagy and polyphagy. Consequently some oligophagous species eat more plant species than some polyphagous ones. This makes sense provided we bear in mind that oligophagy and polyphagy do not refer just to the number of plants eaten, but are intended to convey some information about the selection behavior of the insect.

Polyphagous insects do not eat every plant they encounter, and this is true even for species that eat a very wide range of food plants. The desert locust, *Schistocerca gregaria*, for example, is recorded as eating over 400 plant species, but some other species are rejected without any feeding, and, even amongst these 400, not all plants are consumed in equal amounts. This wide range of acceptability levels amongst host plants is almost certainly a common feature of polyphagous insects. Other examples of polyphagous insect species are the aphid, *Aphis fabae*, which is recorded as living regularly on plants from 33 genera and occasionally on another 39, and the caterpillar of the moth, *Spodoptera littoralis*, known as the Egyptian cotton leaf worm. This species, from Africa and the Mediterranean, has been recorded feeding on over 100 plant species from 49 families.

It is sometimes found in the laboratory that insects will feed and, perhaps, survive on plants that are not eaten in the field. For example, in northern Europe, the swallowtail butterfly, *Papilio machaon*, has one main host-plant species, *Angelica archangelica* in the family Apiaceae (=Umbelliferae), despite that fact that larval survival in the laboratory is good on a number of other species from the habitat. This restriction of host-plant use is determined by the oviposition behavior of the females and a population that is almost monophagous in the field would certainly be classified as oligophagous in the laboratory. Selection of suitable plants for oviposition by the adult female is important in determining the host-plant range of many insects with relatively immobile larvae. Ecological factors may also limit host range (see Chapter 5).

Different populations within a species may differ in the food they eat. A population may be monophagous, or nearly so, while the species as a whole is oligophagous or polyphagous. For example, Colorado potato beetles in Utah were found exclusively on potato, those in Arizona were on *Solanum elaeagnifolium*, and beetles from New Mexico, were exclusively on *Solanum rostratum*. Each of these populations was monophagous although the species as a whole is oligophagous.

These facts obviously add to the difficulty of classifying feeding patterns. In this book we generally use the terms monophagous, oligophagous and polyphagous to refer to the species, rather than the population, and to refer to what happens in the field, rather than what can occur in the laboratory.

1.2 Patterns of host-plant use

Although examples of monophagous, oligophagous and polyphagous species occur in all the major groups of phytophagous insects they are not uniformly distributed across the different taxa.

The Orthoptera (grasshoppers and katydids) stand apart from all the other orders of insects in their relative lack of host-plant specialization. Almost 60% of grasshoppers that have been critically investigated are polyphagous, and a further 25% are oligophagous on grasses (Fig. 1.1). Proven examples of monophagy are rare, and polyphagy is the predominant feature of grasshopper feeding patterns. This is probably also true for other phytophagous Orthoptera although there is very little data.

In all the other phytophagous insect orders, the ranges of plants eaten by individual species are usually much more limited and 70% or more of the species are oligophagous or monophagous. Fig. 1.1 gives examples from the British fauna. Over 80% of all phytophagous Hemiptera, Diptera, Hymenoptera and Lepidoptera are monophagous or oligophagous, with less than 20% polyphagous. A similar pattern occurs where data are available for other regions and for insect taxa where information is available for the world fauna. For example, all the jumping plant lice, superfamily Psylloidea (order Hemiptera), are monophagous or oligophagous.

Fig. 1.2 shows the approximate worldwide total numbers of phytophagous species amongst the different orders of insects. The greatest numbers occur in the Coleoptera (beetles) and Lepidoptera (butterflies and moths). The Orthoptera, by contrast, comprise only a small proportion of the total. Extrapolation from

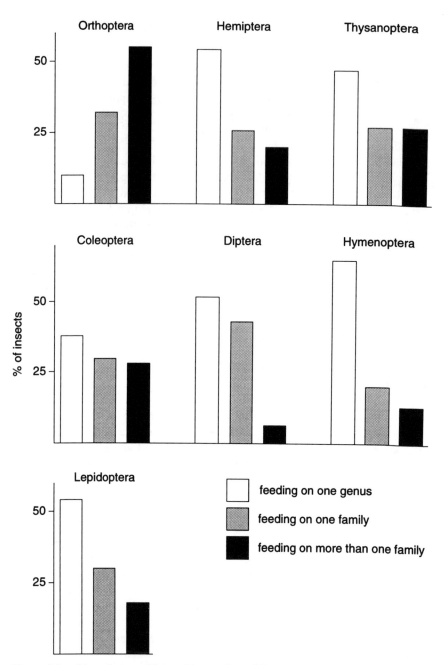

Figure 1.1. Host-plant specificity of insects from different orders, showing the percentage of species that feed on one genus, one family or more than one family. Percentages for Orthoptera are based on all species of grasshoppers for which critical food analyses have been carried out; percentages for all other orders are based on data for the British fauna (from Chapman, 1982).

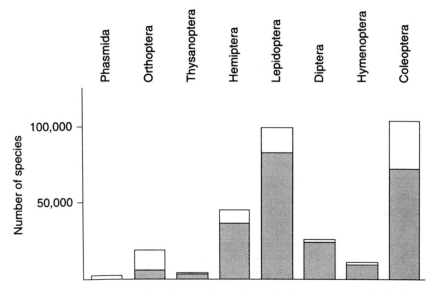

Figure 1.2. Numbers of phytophagous insects in each of the major orders of insects. Cross-hatching shows the numbers that are either monophagous or oligophagous. The vast majority of species have restricted host-plant ranges. Data on specialization within the Phasmida are not available. The degree of specialization is based on data for the British fauna.

the percentages of specialists (monophagous or oligophagous) in the British fauna shows that the vast majority of phytophagous insect species exhibit a high degree of host-plant specificity (Fig. 1.2, shaded areas). Out of approximately 310,000 phytophagous insect species, over 230,000, about 75%, are monophagous or oligophagous.

1.3 Specialization on plant parts

Not only do most phytophagous insect species tend to specialize with respect to the plant taxa they feed on, they also tend to be associated with particular parts of plants. This is at least partly a matter of size and mobility. Grasshoppers, for example, which are relatively large and mobile, usually feed on leaves, but also commonly eat flowers and seeds if these are available, and may sometimes completely destroy small plants. This is also true of some larger caterpillars, like the armyworms, *Spodoptera* species.

In contrast, specialization on particular plant parts is usual amongst the larval forms of holometabolous insects (flies, beetles and Lepidoptera), and especially of those that mine within plant tissues. Sometimes, most members of a major group of insects specialize on a particular organ of a plant; some examples of this are given in Table 1.1. This is not always true, and Fig. 1.3 shows the feeding sites of larvae of different species of weevils in the genus *Apion*. These larvae feed within the tissues, with different species utilizing different plant parts. The adults of all the species usually feed on the leaves and feed on more plant species than the larvae. Fig. 1.3 also illustrates the point that all parts of the plant may be attacked.

In addition to specializing on specific parts of a plant's anatomy, some insects also feed from specific tissues. For example, leaf mining larvae of the moth genus *Phyllocnistus* complete the whole of their development in the leaf epidermis. The highest degree of tissue specialization occurs in the sucking insects of the order Hemiptera. This specialization is possible because their mouthparts form a long and, often, very narrow proboscis. This enables the insects to penetrate the leaf tissues and feed from individual cells in the plant. Some feed primarily from the mesophyll of the leaf. Many aphids and planthoppers feed from phloem, while cercopids feed from xylem (Fig. 1.4). Other species are seed feeders. These insects commonly exhibit morphological and physiological adaptations associated with their particular mode of feeding. As well as being tissue specific, most of these species are also host-species specific.

1.4 Conclusions

This brief account demonstrates that a majority of insects are very specific in the plant species and in the parts of plants that they feed on. In the remainder of this book we address the questions of how insects find their host plants, how they determine that they have reached the appropriate plant or part of a plant, and what may influence their levels of specificity.

Table 1.1. Some insect groups in which a majority of species feed as larvae on particular plant parts.

Plant part	Insect group	Common name
Leaves	Chrysomelidae	leaf beetles
	Gracillariidae	leaf blotch miner moths
	Agromyzidae	leaf miner flies
Stems	Cerambycidae	long horn beetles
Roots	Cicadoidea	cicadas
	Elateroidea	click beetles
Fruits	Dacinae	fruit flies
Seeds	Bruchinae	seed beetles

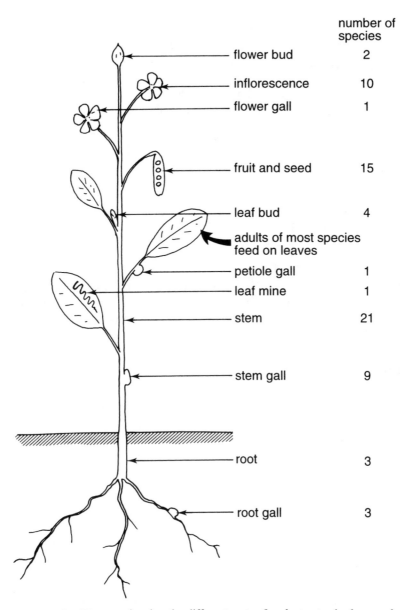

	number of species
flower bud	2
inflorescence	10
flower gall	1
fruit and seed	15
leaf bud	4
adults of most species feed on leaves	
petiole gall	1
leaf mine	1
stem	21
stem gall	9
root	3
root gall	3

Figure 1.3. Diagram showing the different parts of a plant eaten by larvae of species of the weevil genus, *Apion*. Numbers show the number of weevil species with larvae feeding in the part indicated. The figure includes all British species irrespective of their host-plant specificity. All the plant parts would not be affected on one plant species. Adults (thick arrow) generally feed on leaves (data from Morris, 1990).

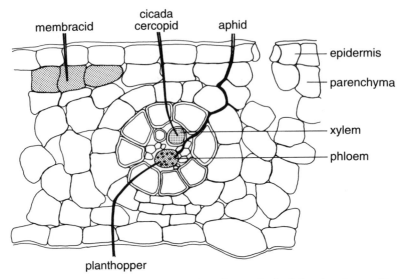

Figure 1.4. Diagrammatic cross-section of part of a leaf showing the specific tissues from which various plant-sucking insects feed. The figure does not imply that all membracids feed from parenchyma, or that all aphids are phloem feeders, although this is often the case.

Further Reading

Eastop, V.F. 1973. Deductions from the present day host plants of aphids and related insects. Symp.R.Entomol.Soc.Lond. 6: 157–178.

Eastop, V.F. 1973. Diversity of the Sternorrhyncha within major climatic zones. Symp.R.Entomol.Soc. Lond. 9: 71–88.

Fox, L.R. and Morrow, P.A. 1981. Specialization: species property or local phenomenon? Science 211: 887–893.

Strong, D.R., Lawton, J.H. and Southwood, R. 1984. *Insects on Plants*. Harvard, Cambridge.

References (* indicates review)

Bernays, E.A., Chapman, R.F., Macdonald, J. and Salter, J.E.R. 1976. The degree of oligophagy in *Locusta migratoria* (L.). Ecol.Entomol. 1: 223–230.

*Chapman, R.F. 1982. Chemoreception: the significance of receptor numbers. Adv.Insect Physiol. 16: 247–356.

Chapman, R.F., Bernays, E.A. and Wyatt, T. 1988. Chemical aspects of host-plant specificity in three *Larrea*-feeding grasshoppers. J.Chem.Ecol. 14: 561–579.

Hsiao, T.H. 1978. Host plant adaptations among geographic populations of the Colorado potato beetle. Entomologia Exp.Appl. 24: 437–447.

*Hsiao, T.H. 1988. Host specificity, seasonality and bionomics of Leptinotarsa beetles. *In* Jolivet, P., Petitpierre, E. and Hsiao, T.H. (eds.) *Biology of Chrysomelidae*. Kluwer, pp. 581–599.

Morris, M.G. 1990. Orthocerous weevils. Handbooks for the Identification of British Insects, vol. 5, part 16. Royal Entomological Society of London.

Smiley, J.T. 1978. Plant chemistry and the evolution of host specificity. New evidence from *Heliconius* and *Passiflora*. Science 201: 745–747.

Wiklund, C. 1982. Generalist versus specialist utilization of host plants among butterflies. *In* Visser, J.H. and Minks, A.K. (eds.) *Insect-Plant Relationships*. Pudoc, Wageningen, pp. 181–191.

2

Chemicals in Plants

The acceptance or rejection of plants by phytophagous insects depends on their behavioral responses to plant features. These features may be physical or chemical. Morphological characters of plants can influence acceptability, either directly by providing suitable visual cues, or by influencing the ability of insects to walk on or bite into tissue. Furthermore, most species of phytophagous insects are confined to certain plant parts, and this will determine the physical and chemical attributes to which the insects respond. The more detailed anatomy and its associated chemistry may constrain or otherwise influence feeding of small insects in particular.

Differences in concentration of nutrients in plants influences host selection. However, it is clear (see Chapter 4) that the narrow host ranges of phytophagous insects most commonly depend on the presence or absence of a variety of secondary metabolites in plants, so that plant chemotaxonomy is probably an important factor in understanding host ranges.

The first sections in this chapter are arranged with respect to the order in which an insect might encounter plant-related chemical stimuli, starting with volatiles detected at a distance, then dealing with the plant surface, and, lastly, the features internal to the plant.

2.1 The first chemicals detected: volatiles

All plants release volatile molecules including water vapor. When the stomata are open, many chemicals are released, especially during active growth. To a large extent the plant probably has little control over this loss, since it will be linked to water loss and regulation of stomatal opening. The volatile chemicals include a wide variety of short chain alcohols, aldehydes, ketones, esters, aromatic phenols, and lactones, as well as mono- and sesquiterpenes. Green plants produce a series of volatile components as a result of the metabolism of lipids. These are commonly referred to as "green odor volatiles." The most important include a variety of 6-carbon alcohols and aldehydes formed by oxidative degradation of leaf lipids (Table 2.1).

In addition to the "green odor" chemicals, many plants have characteristic chemicals. For example, citral is the odor of lemon, nona–2,6-dienal of cucumber and l-carvone of spearmint. Citral and carvone are monoterpenes and many of these terpenoid compounds are produced in quantity from special glands, including glandular trichomes. Isothiocyanates characterize various members of the Brassicaceae (=Cruciferae). These are breakdown products of glucosinolates (see section 2.4.5). The glucosinolates are sulfur-containing compounds which incorporate glucose (and are thus glucosides). During active growth release of the volatile isothiocyanates (i.e., the nonglucose part, or aglycone) may be very high and the total production of the compounds represents as much as 0.7% of the total growth in milligrams dry weight per day. Commonly the characteristic odor of a plant is produced by a group of chemically-related volatiles, such as groups of terpenoids in conifers (e.g., citral, caryophyllene, camphor, citronellal) or mixtures of sulfides from onion and garlic. Analyses of the air surrounding plants, the so-called headspace, yield mixtures of volatiles that may number dozens or even a hundred compounds. For example, the headspace odor of corn silk (*Zea mays*), which is attractive to certain moths, contains 30 compounds, and 40 compounds have been identified in the headspace of sunflower.

The smaller molecular weight compounds tend to be the most volatile, and reach measurable concentrations at greater distances from the plant. As a result, these are of most importance in attraction to the host plant from a distance. At close range, more complex mixtures, including compounds of high molecular weight, predominate, and the concentrations may be very high, especially in the boundary layer very close to the plant surface. This boundary layer is produced by frictional drag, and it varies in thickness, depending on windspeed and leaf-surface morphology, but never exceeds a few mm. Here, even long-chain esters in the wax (e.g., 36-carbon), which are only slightly volatile, can also become part of the aerial bouquet.

Many factors influence the amounts and the profiles of odor components

Table 2.1.   Examples of common green odor components of foliage.

Alcohols	Aldehydes	Ketones and Esters
1-hexanol	hexanal	3-pentanone
cis-3-hexenol	propanal	butanone
trans-2-hexenol	butanal	4-heptanone
1-hexen-3-ol	trans-2-hexenal	cis-hexenyl acetate
1-octanol	cis-3-hexenal	
2-heptanol		

emitted by plants. For example among cultivated mints and lavender, it is known that the characteristic monoterpenes build up during the day and the aroma from plantations is greatest late in the afternoon. Environmental factors that cause closure of plant stomata, such as water stress, reduce the emission of certain volatiles from within the leaf, while damage by more extreme stress or by pollutants increases release of volatiles, partly by causing cell damage and partly because of stomate opening.

2.2 Surface waxes and other surface compounds

Plant surface waxes are complex mixtures of fatty acids, esters, alkanes and other hydrocarbons, and also contain varying quantities of many different secondary metabolites which dissolve in the wax. Generally, the wax layer is embedded in the other cuticular layers, and coats the entire surface. The outer wax layer varies in thickness from little more than a monolayer in aquatic plants to conspicuously thick layers in some others. An inner amorphous layer frequently supports an outer crystalline layer, which gives some plant surfaces their very white waxy appearance known as a "bloom" (Fig. 2.1).

Most plant species have characteristic blends of wax components, providing the potential for host-plant selection by phytophagous insects. It is known that waxes play a major role at least in some cases. The principal classes of constituents are shown in Table 2.2. Alkanes are universally present. These are saturated, long-chain hydrocarbons usually C_{17} to C_{35}, with odd-chain numbers predominating. They are water-repellent and protective, and they may occur as oxidized products, such as alcohols, aldehydes and ketones. Esters of the alcohols and

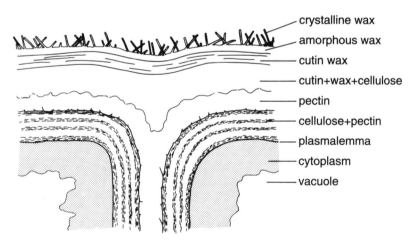

Figure 2.1. Diagrammatic cross section of plant cuticle showing wax layers and parts of the epidermal cells (after Juniper and Jeffree, 1983).

Table 2.2. Principal classes of constituents of plant epicuticular waxes (after Baker, 1982).

Class	Chain length	Common constituent
Odd carbon number		
Hydrocarbons	C17-C35	nonacosane, hentriacontane
Alcohols	C21-C33	nonacosanol, hentriacontanol
Ketones	C23-C33	nonacosanone
β-diketones	C29-C33	hentriacontane-14,16-dione
Even carbon number		
Primary alcohols	C16-C32	hexacosanol
Aldehydes	C22-C32	hexacosanal
Fatty acids	C12-C32	hexadecanoic acid
Esters	C28-C72	hexacosyl hexadecanoic
Triterpenoids		ursolic acid, β-amyrin

long-chain fatty acids also occur. Many other compounds occur in wax, including terpenes and other lipid-soluble secondary metabolites. Some plants having glandular trichomes release their products onto the leaf surface, and in this way the complexity of the surface chemistry is increased. In some plant species there are layers of more polar compounds including phenolics. Water-soluble secondary metabolites have also been detected. In a few cases, flavonoid glycosides are the major chemicals on the leaf surface. Certain nutrients leach out from the cells onto the leaf surface and occur in low but varying levels, with the amounts dependent on leaf age, damage, temperature, insolation, rain and dew. Carbohydrates (mainly sugars) may reach concentrations of 10^{-5}M on moist leaf surfaces. Highest concentrations tend to be over veins and so could potentially influence microsite selection by very small insects.

In addition to the chemistry of the leaf surface wax, resulting from the expression of genes within the plant, a variety of organisms can influence the chemical environment of the leaf surface, or phylloplane. These are mainly bacteria and fungi and they commonly occur at densities of around 10^5 cm^{-2}. In addition, pollutants such as acid rain can alter the surface chemistry and surface-living microflora.

The morphology and composition of plant waxes changes during development of plants. Early leaves of wheat and barley produce mainly primary alcohols, while later leaves produce wax rich in β-diketones. In seedling sorghum, shorter chain lengths of alkanes and esters predominate, while in older plants, longer chain lengths are more abundant. In addition, the seedlings have very high levels of p-hydroxybenzaldehyde.

In many plant species there are genetic differences in the quantity or constituents of the wax. For example on pea leaves over 50% of the wax is the alkane hentriacontane, but four mutants are known with consistently less of the alkane and more of the alcohols. In cabbages, S.Eigenbrode found that wax amounts and chemical profiles in different cultivars had marked effects on susceptibility to the larvae of diamondback moth (*Plutella xylostella*) (see section 4.2.3). However, the precise wax composition is the product of the interaction between the environment and the genetic makeup of the plant. High light levels, low humidities, wind, and high temperatures all induce the production of thick layers of wax.

2.3 Internal components: nutrients

Plants and the parts of a plant vary considerably in their nutritional value for insects. This contrasts with animal tissue and has important consequences for insect herbivores and their selection of food. The reproductive parts of plants are generally nutritious because of relatively high levels of protein. However, they often also contain high levels of secondary metabolites that influence acceptability. A major constraint in use of these tissues is the short-lived nature of flowers and fruit. Leaves usually provide the greatest biomass of the plant and, next to reproductive parts, the best food nutritionally. However, as explained in more detail below, both nutrients and the variety of secondary metabolites are subject to change with development, and fluctuate as a consequence of environmental factors. Stems and petioles, as transport and connecting organs, tend to be low in protein because of high levels of structural materials such as lignin. In stems the transport tissues are dominant, and consequently the sugar levels may be relatively high and secondary metabolites low. Roots are the permanently available foods in many perennials. They tend not to have high levels of nutrients or secondary metabolites, though they can, at the end of a growing season, have very high levels of storage nutrients. They also have different profiles of secondary metabolites from the rest of the plant. They tend to have relatively few physical barriers to invasion.

2.3.1 Proteins and amino acids

Protein is the major nutrient required by phytophagous insects, and is most commonly the limiting nutrient for optimal growth of insects. Plants contain many different proteins which vary in their value for insect herbivores. The most valuable tend to be the soluble enzymes which are easy to extract and digest. Ribulose bisphosphate carboxylase, which fixes carbon during photosynthesis, may make up 50% of the soluble protein in young leaves, and so may be especially important to herbivorous insects. In addition, the overall food quality, in terms of the balance of essential amino acids in the available protein, varies from species to species, and in different plant parts. Among varieties of lettuce for example, it was found that in the variety, romaine, the protein mixture was exceptionally good for growth of many species of grasshopper, while other varieties were poorer due, in part, to relatively low levels of aromatic amino acids.

Free amino acids only make up about 5% of the nutrient nitrogen in plants. Their overall concentration is not necessarily correlated with protein content, but because they can be detected by many insects they may have a more important role in food selection than protein. They vary in concentration with plant part, plant age, and with various abiotic factors. Glutamic acid, glutamate and asparagine are usually the most concentrated. There are chemotaxonomic differences also. For example, members of the Rosaceae are exceptionally rich in arginine. Amino acids are probably of major significance to phloem feeders where the highest concentrations occur. Examination of aphid feeding and growth with respect to amino acids in the phloem show that levels of the more concentrated ones, asparagine and glutamine, are usually positively correlated with insect performance, while there are no consistent patterns with the others.

The major differences in nutritional nitrogen of different leaf tissues are shown in Fig. 2.2. Proteins and amino acids make up about 1% to 40% of the dry weight of leaves. The protein levels of mature leaves tend to differ among families, with Ericaceae and Myrtaceae being exceptionally low; most woody plants and grasses are also low. Highest levels of leaf protein occur in herbaceous legumes.

Variation in protein concentration is observed in relation to taxon, age, and soil nutrient status. It tends to be higher in reproductive parts, and lower in stems. In most plants nitrogen levels are higher in younger tissue than in older leaves, and the levels decline overall as the plant matures. In a temperate climate, this means that the levels decline over the growing season (Fig. 2.3a). Further examples may be found in a review by Slansky and Scriber (1984). In addition, there is a decrease in the proportion of soluble protein, since a larger fraction will be associated with structures such as thicker cell walls, from which extraction is difficult. Availability of soil nitrogen is another major factor in individual plant variation, as demonstrated by the increase in protein levels following fertilization (Fig. 2.3b). Legumes, with their associated nitrogen-fixing bacteria, usually have adequate supplies for proteins as well as nitrogen-containing deterrents and toxins.

Drought causes an accumulation of amino acids, notably proline, as well as cyclitols, betaine and inorganic ions. These effects may maintain the osmotic balance within the plant. In plants that are drought-resistant, higher levels of proline accumulate, and levels as high as 1.2 mg/g dry wt have been recorded in water-stressed Bermuda grass, *Cynodon dactylon*. In legumes, proline is replaced by the cyclitol, pinnitol, as an osmotic effector during drought. The effect of SO_2 pollution is to increase levels of sulfur amino acids and the tripeptide glutathione. This can be beneficial to insects as it is common for plant proteins to have low levels of essential sulfur amino acids.

Free amino acids are of particular importance to phloem feeders. They increase

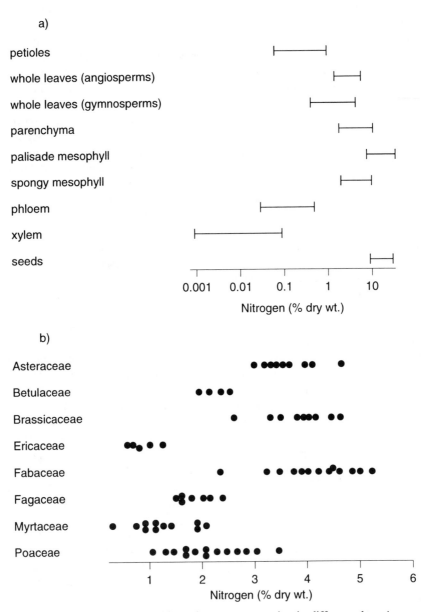

Figure 2.2. **a)** Ranges of nutrient nitogen concentration in different plant tissues (after Mattson, 1980). **b)** Approximate midsummer levels of nutrient nitrogen in leaves of plants in different families. Each point represents the value for one species (data from various authors).

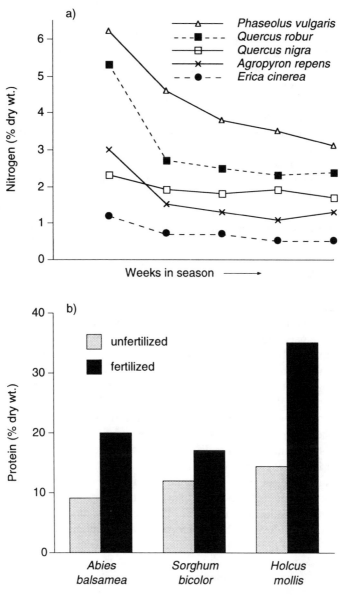

Figure 2.3. **a)** Change in levels of leaf nitrogen of five different plant species through a temperate region growing season. The time span involved was divided into four sectors, since the actual times are different in each case (after Bernays, unpublished; Feeny, 1970; Faeth et al.,1981; McNeill and Prestige, 1982). **b)** Effects of fertilization on leaf protein levels (as a % of dry weight) in three plant species (after Prestige, 1982; Shaw and Little, 1972; Bernays, unpublished).

in phloem in response to changes in plant physiology that involve transport needs. It has been shown that amino acids are present in high concentrations in trees during spring, as they are in herbaceous plants just before flowering, and in senescing leaves.

Xylem has extremely low levels of nitrogen and there are two major plant types. Some plants, such as cotton, transport nitrate from the roots to leaves in the xylem and nitrate is converted to amino acids in the leaves. There are no xylem-feeding insects associated with these plants. Other species of plants, including many grasses and legumes, reduce nitrate in the roots and therefore transport nitrogen from the roots as glutamine. Among trees, reduction of nitrate most commonly occurs in the roots, but the organic solutes of nitrogen that are transported up in the xylem vary. These solutes are generally in the form of arginine and asparagine, but occur in various other forms, such as citrulline in *Alnus*, and glutamine in *Pinus*.

2.3.2 Carbohydrates

Carbohydrates are important nutrients for phytophagous insects. In plants, they may be in the form of starches (glucose polymers), fructosans (fructose polymers) and sugars. Sugars, are of greatest importance and are universal phagostimulants (see Chapter 4). The disaccharide sucrose, the major sugar in plants, varies considerably in concentration (see below), while the hexose sugars are at lower concentrations and vary less. Some plants, such as members of the Poaceae (=Gramineae), tend to have high levels of soluble carbohydrates. There are also plant chemotaxonomic differences in the presence of various sugars. Among sugar alcohols, for example, sorbitol is characteristic of the family Rosaceae, pinnitol of the family Caryophyllaceae, dulcitol of Celastraceae, and quercitol is characteristic of the genus *Quercus*.

Hexoses and sucrose occur in cytoplasm and especially in photosynthetically active tissue. Sucrose is the usual carbohydrate transported around the plant from the point of its production in actively photosynthesizing leaf tissue. For this reason, it is at high level in the phloem, and the bulk of the sucrose in a normal green leaf is in the phloem. In a few families, oligosaccharides are the phloem sugars. For example, in Cucurbitaceae stachyose is the main transport sugar. In some trees, the sugar alcohols mannitol and sorbitol are the translocated carbohydrates.

Starch, a glucose polymer, is the usual plant storage carbohydrate and is a phagostimulant for some insect species. It is the storage carbohydrate material of most plants and occurs in granules associated with chloroplasts or in special storage cells in photosynthesizing tissue. Concentration commonly ranges from 5 to 8% of the leaf dry weight. In grasses having a Hatch/Slack photosynthetic pathway (C4), concentrations up to 10% have been recorded. A few plant groups store polymers of fructose. Inulin, for example, is stored in some composites but is not detected or utilized by insects. Grasses with just the Calvin photosynthetic pathway (C3) accumulate fructosans during cold periods to levels of up to 20% dry weight.

With respect to digestible carbohydrates, there is differential distribution of sugars and polysaccharides. Within leaves, starch is mainly accumulated in granules in the photosynthetic tissue, while sucrose is to a large extent contained in the phloem. However, sugars can reach high levels in the cytoplasm of some plants under freezing conditions, when it is assumed that the sugars provide some tolerance.

Sugar concentrations vary considerably in relation to levels of light and typically increase during the day and decline at night when there is conversion of sugar to starch. Concentrations are higher on sunny days than on cloudy days. Superimposed on this are seasonal changes, and differences depending on the degree of insolation received by different parts of the plant (Fig. 2.4). This variation may be very significant for insects. For example, those feeding at the end of the day or after dark will maximize the amounts of sugar in the diet; those that choose the sunny side of trees for thermoregulatory reasons will automatically encounter greater sugar concentrations in their food. An increase in soluble sugars in plants is a general response to the pollutants SO_2 and O_3. This could be one of the reasons for improved performance of some herbivores under conditions of pollution by these gases.

It is very common for protein and sugar levels to be inversely correlated in leaves (Fig. 2.5). The selection of plants with high sugar contents by insects will then reduce the relative intake of protein. In addition, younger leaves are usually richer in protein and older leaves tend to be richer in carbohydrates. Specialist insects selecting among leaves of their host plants are potentially able to obtain a balance of carbohydrate and protein by feeding on different aged leaves; such mixtures have been noted in observation of *Manduca sexta* feeding on tobacco plants.

2.3.3 Lipids and minor nutrients

Plants contain a wide variety of lipids including triglycerides, phospholipids, glycolipids, steroids and the multitude of components that make up the surface wax. Many different phytosterols have been identified with sitosterol being the most widespread and abundant. However, it appears that many species in the family Chenopodiaceae have rather low levels of sitosterol and higher levels of spinasterol, stigmasterol and brassicasterol. Seed oils and other stored lipids vary with taxon, so they may have some significance for host affiliation of seed-feeding insects though this is not known.

Most leaf tissue has an acid pH due to the presence of various organic acids. Most of these are components of the tricarboxylic acid cycle, and some of them, especially citric and malic acids, accumulate to levels as high as 5% of the dry

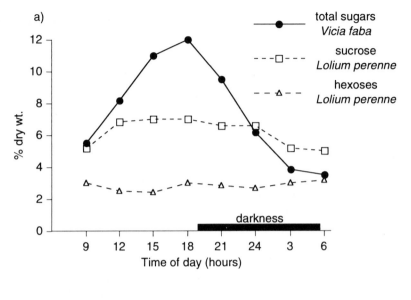

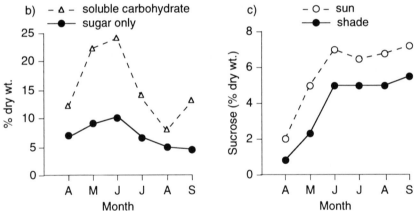

Figure 2.4. **a)** Changes in sugar levels during the day in the bean, *Vicia faba*, and a grass, *Lolium perenne* (after Waite and Boyd, 1953; Cull and van Emden, 1977). **b)** Changes in sugar levels in the grass, *Agropyron repens*, during the growing season in England, 1976 (after Bernays unpublished). **c)** Changes in sugar levels in the oak, *Quercus robur*, during the growing season in England, 1969 (after Feeny, 1970).

weight. Others include formic acid, which occurs in stinging nettle hairs, acetic acid, tartaric acid and oxalic acid. The latter commonly occurs as a crystalline calcium salt (Fig. 2.6g). The crystals exhibit a range of forms which are of taxonomic significance. Needle-shaped crystals, or raphides (Fig. 2.6b), are common. These occur in many plants, but in large amounts in *Oxalis* and a

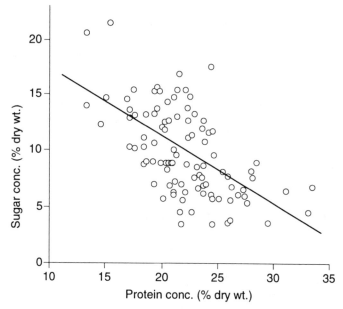

Figure 2.5. Relationship between protein and sugar concentration in leaves of ragwort, *Senecio jacobaea*. Each point shows the values in the leaves of one plant (after Soldaat, 1991).

number of other species, especially in the family Onagraceae. Ascorbic acid is universally present in plants and essential for most herbivorous insects. It is found at high concentrations in young tissue. Commonly occurring organic acids also include longer chain acids, such as linoleic and linolenic acids (see Fig. 2.28) which are nutrients for insects, although many of the carboxylic acids are also utilizable by insects. Phenolic acids add to the acidity of leaf tissue and are discussed in section 2.4.2.

Minor nutrients, including inorganic salts, vary greatly in plants according to taxon and soil contents. Herbaceous plants have generally high levels of mineral nutrients (about 10% dry weight) compared with woody plants. Potassium is at much higher levels than sodium in most plants and is particularly high in some families, for example, the Boraginaceae. Some members of the Chenopodiaceae accumulate sodium. For example, in saltbush, *Atriplex* spp., NaCl may represent more than 10% of the dry weight. Soil type affects mineral composition of plants considerably. For example plants on serpentine soils tend to be high in magnesium.

2.4 Internal components: secondary metabolites

The nutrient compounds in plants, and all the additional substances considered "essential for plant growth and development" are together sometimes referred to as primary metabolites. The enormous array of compounds which appear not to be so essential for the basic growth of plants, and for which functions are not always known, are commonly referred to as secondary metabolites. Many are biologically active compounds, being known poisons for animals, fungi or micro-organisms. Others may have different ecological functions, such as aiding in competitive interactions against other plants, or providing protection from abiotic factors. No two plant species have the same profiles of secondary metabolites, and thus many species may be identified by their chemistry. Consequently, they are of profound importance in the selection of hosts by phytophagous insects. Some authors refer to them as plant allelochemics.

Many secondary compounds present in plants are toxic, not only to potential herbivores, but also to the plant itself. For this reason, they are usually either compartmentalized and separated from cytoplasm, or they are stored in an inactive form. Anatomy is of particular importance in relation to the precise location of secondary metabolites in plants. Some alkaloids are sequestered in epidermal tissue vacuoles or in latex, others are to be found only in vacuoles of young tissue. Some, such as nicotine are manufactured in the roots and transported to the aerial parts; they are at high level in xylem. Others are deposited in cell walls, and in trees may then end up in the bark. Nonprotein amino acids are usually in highest concentration in seeds. Coumarins tend to be localized in oil glands or in cells of the epidermis. Acetylenes and other lipid-soluble compounds may be secreted into the surface wax.

Terpenoids are always sequestered in specific sites. In laurel leaves, single cells are modified and become nonliving cells that contain monoterpenes; other plants such as cotton have secretory pockets in which the sesquiterpenes accumulate; the resin ducts of pine needles contain mono-, sesqui- and diterpenes; the laticifers of *Euphorbia* spp. contain triterpenes. Many species have glandular trichomes in which diverse terpenoids and phenols are sequestered. Some examples of microdistribution of chemicals are shown in Fig. 2.6 and Table 2.3. The plants, then, are mosaics of tissues with high concentrations of particular secondary chemicals and tissues with none. The implications for herbivory are important, since larger species will automatically ingest them in a mouthful of food and will taste the average concentration. Smaller species may either avoid them or, if they encounter the chemicals, the concentrations will be extremely high. Mining insects avoid surfaces, glands or vessels that are noxious. Phloem and xylem suckers also avoid vacuoles and, if they penetrate between cells, they avoid the metabolites produced enzymically by damage to the plant. On the other hand they encounter chemicals deposited in cell walls.

A very common way in which plants store quantities of secondary metabolites, is to chemically combine them with sugars, salts or proteins to produce relatively innocuous compounds. When tissues are damaged, the free secondary metabolites are released by enzymic action or oxidation, so that herbivores or pathogens may be faced with concentrated doses just in the area of damage. Details of

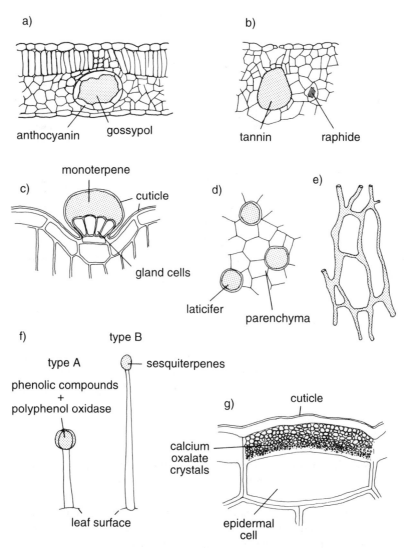

a)

anthocyanin   gossypol

b)

tannin   raphide

monoterpene

c)   cuticle

gland cells

d)

e)

laticifer
        parenchyma

f)   type B

type A   sesquiterpenes

phenolic compounds
        +
polyphenol oxidase

g)   cuticle

calcium
oxalate
crystals

leaf surface   epidermal
                        cell

Figure 2.6. Storage of secondary and other compounds in plant tissues. **a)** Section of a cotton leaf showing a gossypol gland. **b)** Section of a leaf of *Oscularia* showing a tannin-containing cell and bundle of calcium oxalate crystals (raphides). **c)** Gland on the leaf of *Thymus* containing monoterpenes. The gland cells secrete their products beneath the cuticle forcing it into a dome-shaped structure. **d)** Transverse section of a leaf with laticifers. **e)** Anastomosing laticifers in the leaf of *Sonchus*. **f)** Glandular trichomes on the leaf of potato. Type A has a tetralobate, membrane-bound gland. Type B continuously exudes sesquiterpenes from the tip. **g)** Calcium oxalate incrustation in the outer epidermal wall of a leaf of *Conophytum*.

Table 2.3. Examples of secondary metabolites associated with specific tissues or structures.

Tissue	Chemical	Plants
Trichomes	monoterpenes	Lamiaceae
Trichomes	sesquiterpenes	Solanaceae
Trichomes	flavonoid glycosides	Solanaceae
Wax	triterpenes	Asclepiadaceae
Wax	phenolics	Rosaceae
Epidermis	quinolizidine alkaloids	*Genista*
Cell walls	tannins	many trees
Dead cells	tannins	woody plants
Vacuoles	cyanogenic glycosides	many
Vacuoles	alkaloids	many
Oil glands	furanocoumarins	*Citrus*
Oil glands	sesquiterpenes	*Gossypium*
Latex	di- and triterpenes	Euphorbiaceae
Latex	sesquiterpene lactones	Asteraceae
Latex	alkaloids	Euphorbiaceae
Resin canals	diterpenes	gymnosperms
Seeds	non-protein amino acids	legumes
Seed coat	furanocoumarins	*Pastinaca*
Bark	quinine	*Cinchona*

some of these processes are described under the relevant sections below. The compartmentalization of reactants in these processes is important; the conjugated chemical is commonly in the cell vacuole and the appropriate enzymes are in the cytoplasm.

2.4.1 Nitrogen-containing compounds.

Plants producing significant amounts of nitrogenous secondary metabolites tend to be those in which nitrogen is normally plentiful. Taxa that characteristically utilize richer soils, or that have associations with nitrogen-fixing bacteria commonly produce them. They include amines, alkaloids, nonprotein amino acids and cyanogenic compounds. All are known to influence one or more insect species and are probably of considerable significance in host-plant selection, especially since they vary greatly with plant taxon, and include a great variety of chemotypes. Effects of many such compounds are discussed in Chapter 4.

2.4.1.1 Nonprotein amino acids

These are mostly analogues of various protein amino acids. There are approximately 400 known structures. The nonprotein amino acids occur especially in legumes, and are at highest concentrations in seeds, where they are important as nitrogen storage materials. Some are characteristic of certain taxa. For example,

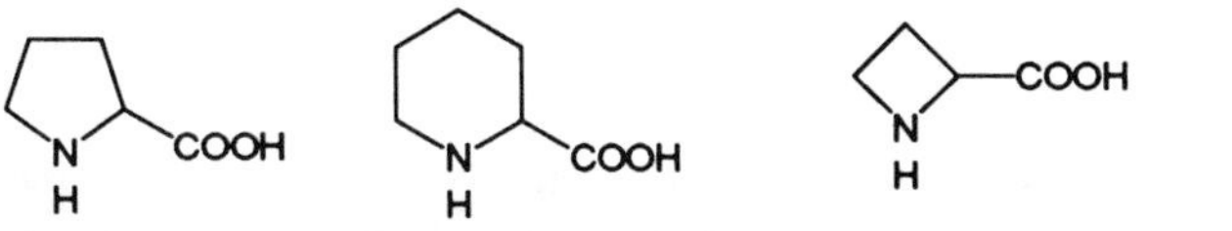

Proline   Pipecolic acid   2-Azetidinecarboxylic acid

[protein amino acid]   [nonprotein analogues]

Figure 2.7.   Structure of proline, a protein amino acid, and two nonprotein amino acid analogues. Some other protein amino acids have similar analogues.

canavanine, an analogue of arginine, occurs only in the Fabaceae. Structures of two proline analogues are shown in Fig. 2.7. Plants can accumulate impressive levels of some nonprotein amino acids, up to 5% of the dry weight of leaves and 10% of the dry weight of seeds. They are often toxic to herbivores because they become incorporated into protein in place of the analogous protein amino acid. This alters the properties of the protein.

2.4.1.2 Amines and alkaloids

Amines are produced when amino acids are decarboxylated (Fig. 2.8). There are about 100 structures known. Some are very active compounds in mammalian physiology, such as noradrenaline which occurs in banana and mescaline from the peyote cactus flowers. Tyramine is very widespread in plants, while histamine and serotonin are more restricted. Hordenine in barley is one of the few known to be an important deterrent to some insects.

Alkaloids form one of largest classes of secondary substances in plants. This group is an extremely heterogeneous array of unrelated classes of biologically active compounds. They are usually defined as "those organic basic compounds which contain one or more nitrogen atoms, usually in combination as part of a

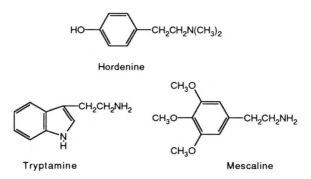

Hordenine

Tryptamine   Mescaline

Figure 2.8.   Structures of three amines.

cyclic system." Over 6,500 alkaloids are known, and some of the variety of structures may be seen in Fig. 2.9.

Alkaloids in concentrations of more than 0.01% dry weight occur in at least 15% of vascular plants. Their occurrence is very uneven in the plant kingdom but some families are very rich in certain groups, and some taxa have characteristic groups. In no case, however, is a class of alkaloids confined to any plant taxon. For example, plants in the tribe Genisteae, of the family Fabaceae characteristically contain quinolizidine alkaloids, although these compounds do occur elsewhere. Pyrrolizidine alkaloids are most common in the family Boraginaceae, but there are sporadic examples in several other families. Table 2.4 lists a variety of plant families rich in alkaloids, the alkaloids, and their chemical classes. Although alkaloids are considered the epitome of plant defence compounds, and do have diverse effects on insect behavior and physiology, they are generally

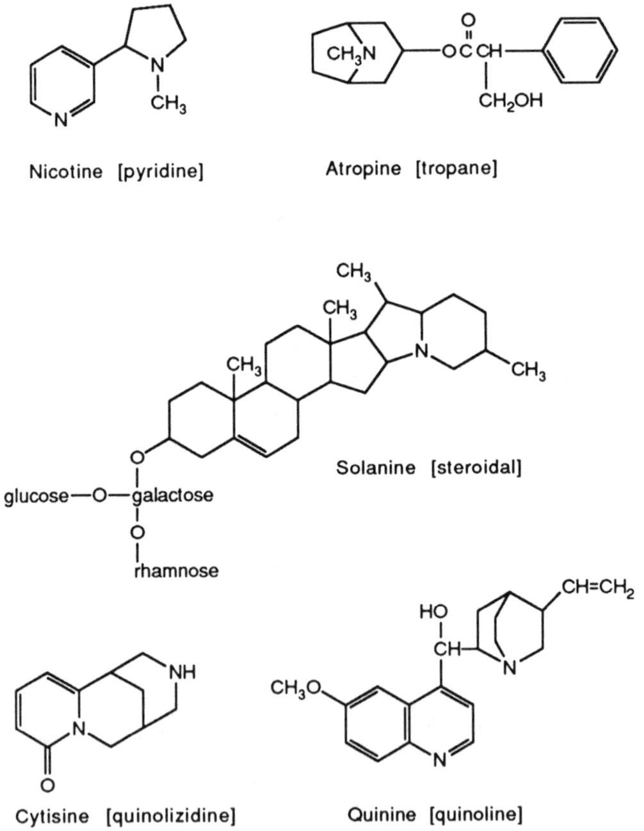
Nicotine [pyridine]
Atropine [tropane]
Solanine [steroidal]
glucose—O—galactose
rhamnose
Cytisine [quinolizidine]
Quinine [quinoline]

Figure 2.9. Structures of five alkaloids with their chemical types in brackets.

Table 2.4. Examples of plant families which tend to be rich in particular groups of alkaloids. It should be noted that the different classes of alkaloids are also widespread in other families of plants.

Family	Species example	Alkaloid	Type of alkaloid
Asteraceae	*Senecio jacobaea*	senecionine	pyrrolizidine
Fabaceae	*Cytisus laburnum*	cytisine	quinolizidine
Liliaceae	*Colchicum autumnale*	colchicine	tropolone
Loganiaceae	*Strychnos nux-vomica*	strychine	indole
Papaveraceae	*Papaver somniferum*	morphine	morphine
Rubiaceae	*Cinchona officinalis*	quinine	quinoline
Solanaceae	*Solanum tuberosum*	solanine	steroidal
	Atropa belladonna	atropine	tropane
	Nicotiana tabacum	nicotine	pyridine

much more toxic to vertebrates. Often this difference is by as much as two orders of magnitude. For humans, most alkaloids taste bitter, but their effects on insect taste responses vary considerably.

2.4.1.3 Cyanogenic glycosides

Approximately 2,500 plant species across many families produce hydrogen cyanide (HCN) in more than trace amounts. HCN is a general respiratory poison, and it is stored in plants in some nontoxic form, often combined with a sugar to form a cyanogenic glycoside (Fig. 2.10). About 30 different cyanogenic glycosides are known and examples of glycosides found in different plants are

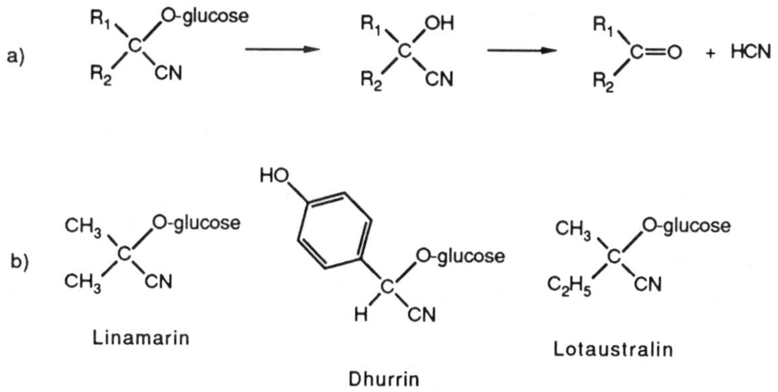

Figure 2.10. **a)** Release of hydrogen cyanide (HCN) by enzymic action on a generalized cyanogenic glycoside. The first arrow represents enzymic hydrolysis; the second arrow represents spontaneous change due to chemical instability of the intermediate product. **b)** Structures of three cyanogenic glycosides. Linamarin and lotaustralin are very widespread in the plant kingdom and generally occur together.

Table 2.5. Examples of plant families and species containing different cyanogenic glycosides.

Family	Species	Cyanogenic glycoside
Caricaceae	*Carica papaya*	prunasin
Euphorbiaceae	*Manihot esculenta*	linamarin
Fabaceae	*Lotus corniculatus*	linamarin, lotaustralin
Fabaceae	*Trifolium repens*	linamarin, lotaustralin
Poaceae	*Sorghum bicolor*	dhurrin
Rosaceae	*Prunus padus*	prunasin
Rosaceae	*Prunus* spp.	amygdalin
Rutaceae	*Zieria laevigata*	zierin

shown in Table 2.5. Both the glycoside and the HCN may affect behavior. Release of the HCN is usually a result of enzyme activity that occurs when the tissue is damaged and the vacuolar glycoside is contacted by the cytoplasmic hydrolysing enzymes. The most common glycosides that produce HCN are linamarin and lotaustralin. The HCN production involves hydrolysis of the glycoside during which the sugar is removed. The remaining moiety is unstable and breaks down to give HCN and an additional compound (Fig. 2.10). The activity of the enzymes is, at least in some circumstances, influenced by the degree of water stress in the plant, and this can have marked effects on host selection if deterrents are released enzymically during chewing (see Chapter 5).

Many cyanogenic compounds have probably not been identified yet and there are, in addition, cyanolipids. These are complexes of cyanide with a variety of lipophilic substances, which are toxic at least to vertebrate herbivores.

2.4.1.4 Betacyanins

These compounds are similar to anthocyanins (see Fig. 2.17, below), but contain nitrogen. They are water-soluble purple pigments, as in beetroot, and provide color for flowers. However they do occur throughout the plant, though the color may not be obvious due to the masking by chlorophyll. They are restricted to the order Centrospermae; families include Cactaceae, Chenopodiaceae, Amaranthaceae and Portulacaceae. In all other orders of plants anthocyanins are produced for flower colors. The distribution of betacyanins in the Centrospermae is one of the clearest cut examples of chemical production associated with a particular plant taxon. Nothing is known about their significance for insect herbivores, however, and they are not known to have toxic properties.

2.4.1.5 Variation in levels of nitrogenous secondary metabolites

In most cases studied to date, concentrations of biologically active N-containing secondary metabolites are highest in young tissue and reproductive parts. Typically, the concentrations in leaves are at their maximum as young tissue becomes vacuolated, and thereafter there is a slow decline to senescence (Fig. 2.11). Within-plant distribution is demonstrated in Fig. 2.12 for quinolizidine alkaloids in lupine, and it is clear that reproductive parts and epidermal tissue have the highest concentrations, and young leaves have higher concentrations than old leaves. This pattern is typical of alkaloids in general. Nonprotein amino acids are at highest levels in seeds, while seedlings tend to have higher levels than older plants. Nitrogen fertilization or shady growing conditions result in higher concentrations of alkaloids, cyanogenic glycosides and nonprotein amino acids in those plants that produce them.

In a number of cases, herbivore damage has been shown to increase the levels of alkaloids in plants. For example, quinolizidine alkaloids in lupines increase after damage by blue butterfly caterpillars, and pyrrolizidine alkaloids in ragworts increase after damage by cinnabar moth caterpillars. Best studied is the case of dramatic increases in nicotine levels in tobacco after feeding by larvae of *Manduca sexta*, the tobacco hornworm. The studies by I. Baldwin demonstrate a fourfold increase in alkaloids after removal of about a third of the leaves. The increase may be seen after a few days but is maximal after about a week (Fig. 2.13). Such changes can be expected to have profound effects on herbivore host choice. Extreme damage may reduce the levels of such compounds.

The presence of endophytic fungi in the plant tissue is a potential source of great variation in the secondary chemistry of the plant, which in turn may influence herbivore host selection. Endophytes are fungi that infect leaves and stems of healthy plants in many different taxa. Commonly, there are no external symptoms of the infection, and almost certainly many endophytes have been

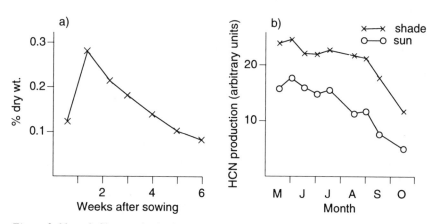

Figure 2.11. **a)** Changes in the level of alkaloids in seedlings of perennial rye grass, *Lolium perenne*, after sowing (after Bernays and Chapman, 1976). **b)** Changes in levels of HCN that can be released from bracken fern, *Pteridium aquilinum*, at different times during growth over the summer in England (after Cooper Driver et al., 1977).

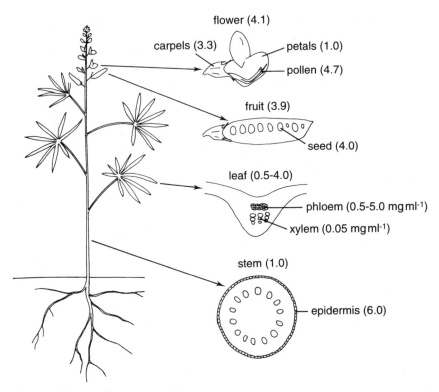

Figure 2.12. Concentrations of quinolizidine alkaloids in different parts of a generalized lupine plant expressed as a percent of dry weight for solid tissues and mg ml^{-1} for xylem and phloem (after Wink, 1992, and references therein).

overlooked. They have been most studied in grasses where they produce a wide variety of alkaloids. Some authors believe that endophytic infections of grasses present a facultative mutualism, by which the plant obtains chemical protection from herbivory, and the fungi are provided with the place to live. Examples of deterrence due to endophytic infections are shown in Table 2.6.

Genetic variation in the production of nitrogenous secondary metabolites is also well established in a few cases. For example, in *Pteridium,* bracken fern, plants may have the cyanogenic glycoside alone, or the cyanogenic glycoside plus a hydrolyzing enzyme that causes release of HCN when leaves are damaged, or they may have neither. The same type of polymorphism has been shown in the family Fabaceae including species of *Trifolium* and *Lotus.*

In the case of lupines, there are genetic variants without the normal quinolizidine alkaloids. These plants are more susceptible to damage by various insects. Less conspicuous genetic variation such as that shown for pyrrolizidine alkaloids in ragwort (Fig. 2.14) is probably common. Other examples include crop plants

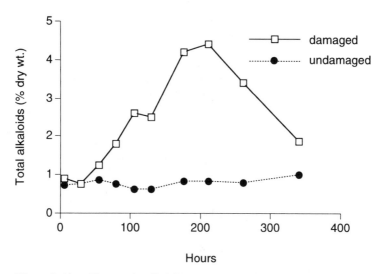

Figure 2.13. Changes in alkaloid content of tobacco leaves at different times after the start of leaf damage by caterpillars. Mean values from 60 plants are shown; caterpillar feeding occurred during the period 0 to 60 hours (after Baldwin, 1989).

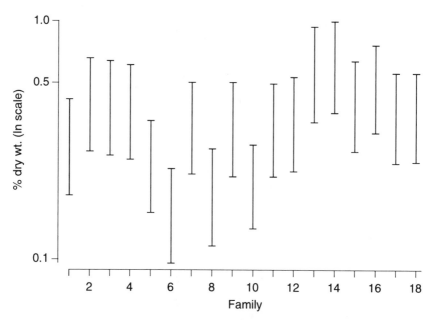

Figure 2.14. Genetic variation in levels of pyrrolizidine alkaloids in ragwort, *Senecio jacobaea*. Eighteen crosses were made from plants in the dune area of The Netherlands. From each cross, two sibs were grown under standard conditions and at 13 weeks, shoots were harvested and analyzed (after Vrieling, 1990).

Table 2.6. *Examples of deterrence of grasses to insects resulting from endophytic infections (after Dahlman et al., 1992).*

Insect species	Common name	Endophyte	Host plants	Behavior affected
Spodoptera frugiperda	fall armyworm	*Balansia*	*Cenchrus, Festuca, Lolium, Paspalum*	feeding
Crambus spp.	webworms	*Acremonium*	*Lolium*	feeding, oviposition
Rhopalosiphum spp.	aphids	*Acremonium*	*Lolium*	settling
Schizaphis spp.	aphids	*Acremonium*	*Festuca*	feeding
Listronotus bonariensis	Argentine stem weevil	*Acremonium*	*Lolium*	feeding, oviposition

in which there has been selection for varieties with low levels of toxic or deterrent compounds. Variation in tomatine content of tomato leaves is controlled by two co-dominant alleles, whereas alkaloid content of potato varieties is under polygenic control.

2.4.2 Phenolics

The common feature of phenolics is an aromatic ring bearing one or more hydroxyl substitutes. They tend to be water soluble. Phenolics are ubiquitous, with numerous different types including simple phenolic acids and their glycosides, flavonoids and their glycosides, coumarins, and polymers of phenols including tannins. Plant species that grow in poor soils tend to produce relatively high concentrations of phenolics. Although this is a constitutive trait, there is additional production of phenolics under the poorest soil conditions. They have various functions in the life of the plant as well as being important in interactions with herbivores. For example, lignin is the structural material of the cell wall, anthocyanins provide the color of flowers, flavonols are important in growth regulation of some plants, phenolic acids and tannins are important in the soil around trees for nutrient cycling, and critically important against pathogens. They are often deterrent to insect herbivores, but many are phagostimulants for particular species. They may be deleterious if ingested, but in some cases are beneficial.

2.4.2.1 Phenols and phenolic acids

There are about 200 different simple phenols. Many simple phenolic acids and their glycosides are very common and widespread. For example, almost universal among the angiosperms are *p*-hydroxybenzoic acid, protocatechuic acid, caffeic acid, and vanillic acid. Others may be more common in one family than in another. For example ferulic acid is relatively high in Poaceae (grasses), ferulic and caffeic acids in monocotyledons generally, and protocatechuic acid in Ericaceae. All of them are very susceptible to change in concentration, increasing greatly with high light levels. Some are precursors of lignin, and are at higher concentration in woody plants. The lower plants, such as mosses, without any lignin, have no phenolic acids either. Some authors subdivide the phenolic acids into cinnamic and benzoic acids (Fig. 2.15).

2.4.2.2 Phenylpropanoids

Phenylpropanoids are defined as having an aromatic ring with a three-carbon side chain and as such could also include the cinnamic acids, discussed in this chapter with the phenolic acids. Chlorogenic acid is the combination of caffeic acid and quinic acid.

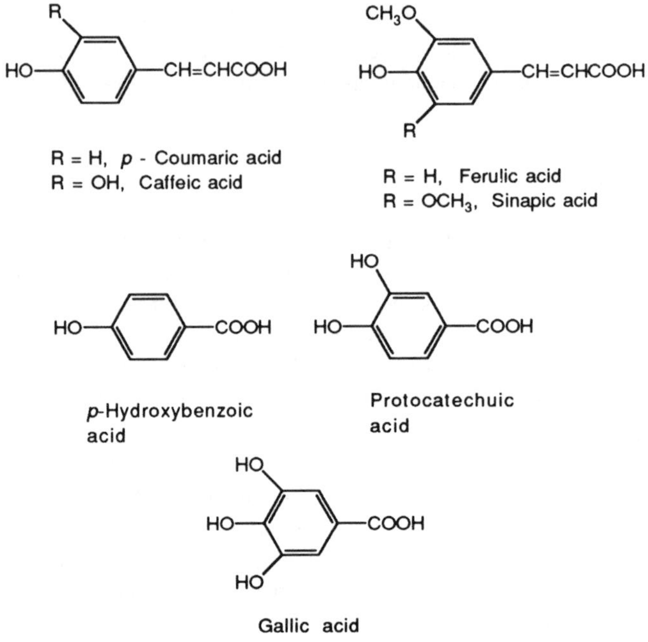

Figure 2.15. Structures of phenolic acids.

The best known phenylpropanoids are coumarins, simple two-ring structures produced from cinnamic acid. The most widespread of these is coumarin itself which occurs in 27 plant families, including Poaceae and Fabaceae. A related structure, umbelliferone, is present in a few families. There are about 50 different hydroxylated coumarins occurring in plants, often as esters or glucosides. The more complex furanocoumarins have a restricted occurrence, being mainly in the Apiaceae (=Umbelliferae) and Rutaceae. They have an additional ring and, depending on where it is attached, are defined as linear furanocoumarins, such as psoralen, or angular furanocoumarins, such as angelicin (Fig. 2.16).

Coumarins have been found in all parts of plants, though roots and seeds have been investigated most thoroughly. Maximum levels have been found in roots where up to 9% dry weight has been recorded. Yields from umbel fruits vary with the stage of maturity but highest levels (1–5% dry weight) occur when fruits are fully formed yet still green. Foliage typically contains about 0.1–1.0% dry weight. Furanocoumarins are often secreted into the surface wax of the leaves where they are presumed to have maximum impact on herbivores. Furthermore, the furanocoumarins are converted into more biologically active materials by sunlight (i.e., they are photoactivated), and some authors consider their position in cuticle to be important in plant defense. Many species of plants containing coumarins have a suite of different types. The profiles of coumarins present,

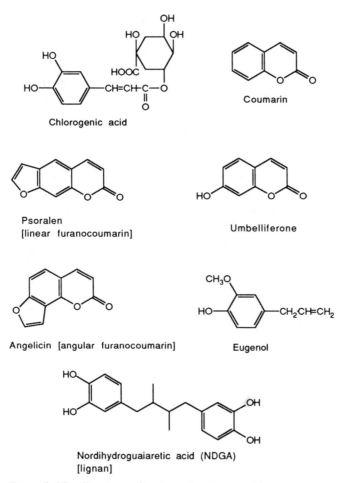

Chlorogenic acid

Coumarin

Psoralen
[linear furanocoumarin]

Umbelliferone

Angelicin [angular furanocoumarin]

Eugenol

Nordihydroguaiaretic acid (NDGA)
[lignan]

Figure 2.16. Structures of various phenylpropanoids.

that is, which ones, and how much of each, may vary more than the total concentrations.

A related group of compounds are the phenylpropenes which are lipid soluble and volatile. They are important in the odors of plants, especially those bearing fruits used as spices. These include eugenol (Fig. 2.16) which is widespread, and at especially high concentration in cloves, and anethole in anise and fennel (Apiaceae). Low concentrations are often important elements of floral odors.

Finally, the lignans are related compounds consisting of two cinnamic-acid-type components condensed together at the side chains. They are widespread and many occur in wood and resin. Two examples are nordihydroguaiaretic acid, from *Larrea* spp. (Fig. 2.16), and sesamin which is widely distributed.

2.4.2.3 Flavonoids, quinones and tannins

Flavonoids may be considered to consist of two benzene rings, A and B, linked by a three-carbon chain. This last may form a pyrone ring (Fig. 2.17). Flavonoids and their glycosides are present in all vascular plants and are extremely diverse, with about 4,000 known structures. They are generally present as glycosides in which a great variety of sugars is used, combined to the flavonoid skeleton in a variety of different ways. Flower color is largely due to the presence of pigments in chromoplasts or cell vacuoles of floral tissues, and anthocyanins are of greatest importance. They contribute orange, yellow, red or blue, and white. They are also present in leaves where their colors tend to be masked by chlorophyll. Different classes are listed in Table 2.7, but only the first three are very widespread. However, the more limited distribution of the last three perhaps gives some possible bases for family-level chemotaxonomy. Isoflavonoids, for example, are only common in the Fabaceae, subfamily Lotoidae. This subfamily includes alfalfa and clover, in which isoflavone structures are superficially similar to certain steroids; some of these may mimic the action of steroid hormones when ingested by mammals.

Quinones are produced from phenols by oxidation as demonstrated in their structures (Fig. 2.18). About 800 occur naturally in plants, especially in bark, heartwood and root. They vary in color from pale yellow to black. They are

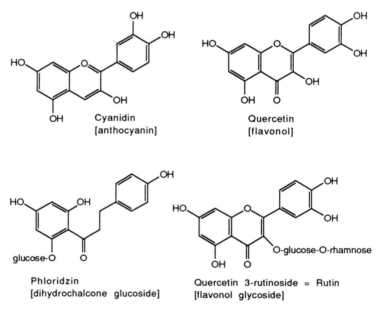

Cyanidin
[anthocyanin]

Quercetin
[flavonol]

Phloridzin
[dihydrochalcone glucoside]

Quercetin 3-rutinoside = Rutin
[flavonol glycoside]

Figure 2.17. Structures of flavonoids. Type of flavonoid is given in brackets.

Table 2.7.   Major classes of flavonoids, their color, occurrence in plant tissues and examples.

Class	Color	Tissue	Example
Anthocyanins	scarlet, red, mauve	petals, leaves, fruits	cyanidin
Leucoanthocyanins	colorless	leaves, wood	building block of condensed tannins
Flavonols, Flavones	colorless, yellow	leaves, stems	quercetin, rutin
Chalcones, Aurones	yellow	flowers	butein, phloridzin
Flavonones	colorless	leaves, fruit	naringin
Isoflavones	colorless	flowers, leaves, roots	rotenone

reactive compounds and many, for example, juglone, are active against germinating seedlings of other plant species as well as against animals. In many plants the presence of polyphenol oxidases is important in the production of quinones. Consequently, in their role in defense against insects, polyphenol oxidase acts on a variety of phenolics to produce quinones when the leaf tissue is chewed and crushed. This is often obvious in the darkening of damaged tissue. These quinones tend to be more deterrent to insects than the unoxidized phenolics, and in some legume crops the varieties producing larger quantities of polyphenol oxidase are more resistant to insect pests.

Tannins are polymers of phenolic compounds, and form a very heterogeneous group. They share the property of being able to bind with protein, and usually precipitate it. There are two broad groups. First, there are the hydrolyzable tannins, which are built up from glucose and phenolic acids especially gallic acid. Second, there are the condensed tannins which are polymerized flavonoids.

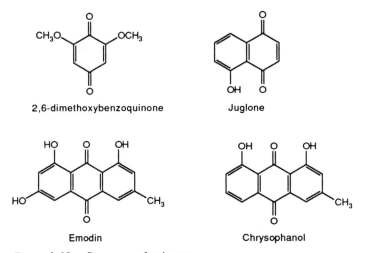

2,6-dimethoxybenzoquinone   Juglone

Emodin   Chrysophanol

Figure 2.18.   Structures of quinones.

The occurrence of tannins is often associated with the woody habit so they are more likely to be found in trees and shrubs than in herbaceous plants.

Sources of commercial tanning materials illustrate the variety of plants and their different parts that have very high tannin concentrations. For example, bark from chestnut, oak, *Eucalyptus*, and *Acacia* spp. is an important source of tannins; heartwood from quebracho and oak yields high levels; seed pods of species of *Caesalpinia* have high levels; and insect galls on leaves of sumac and oak are very rich in tannins.

The major trend in the evolution of the higher plants, with respect to tannins, is thought to be loss, or reduction in amounts produced. The more recent families of plants tend to be herbaceous. In the ferns, 90% of species contain condensed tannins, in gymnosperms 78%, and in angiosperms 36%. Within the dicotyledons, 38% have condensed tannins and 15% have hydrolyzable tannins. In monocotyledons, 24% have condensed tannins and none has hydrolyzable ones. The grass family (Poaceae), is a recent family that contains no hydrolyzable tannins and, only rarely, condensed tannins.

2.4.2.4 Variability in levels of phenolics

Changes in phenolics occur with the stage of growth of particular plants, but the patterns of change are somewhat different in different species. For example, certain tropical trees tend to have high levels of various phenols and flavonoids in young leaves, and the levels fall as the leaves expand. The flavonoids cause the red color of the new leaves of many trees in tropical regions. On the other hand, in temperate regions, trees often show an increase in leaf phenolics, especially tannins, as the leaves mature and senesce. One hypothesis for this latter pattern suggests that the plant is dumping excess carbon from photosynthesis. Another possibility is that tannins improve soil nutrient cycling after leaf fall. There is good evidence for this in the case of trees growing on poor soils.

With the coumarins, that have been most studied in the family Apiaceae (=Umbelliferae), highest levels are found in the seeds, followed by the young leaves and, in some cases, the roots. The presence of furanocoumarins in plants is associated with high light environments and nutrient-rich soil. Ultraviolet light induces biosynthesis of these compounds. On the other hand, acid fog has been shown to increase the levels in celery by over five times. Furanocoumarins also increase after herbivore damage in wild parsnip, and it is probable that such increases occur commonly. Fungal infection of parsley also causes increases in furanocoumarins.

Among phenolics of all kinds, the amounts of light experienced by the plants have the most profound effect on concentrations in the leaves. For example it is not uncommon to see twofold increases with increasing light level (Fig. 2.19). In experiments with sorghum in India, it was found that phenolic levels varied across years; the strongest correlation was with the overall level of insolation.

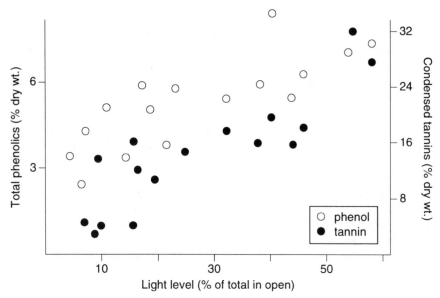

Figure 2.19. Variation in total phenol and condensed tannin in relation to light level, in *Acacia* in Cameroon (after Waterman and Mole, 1989).

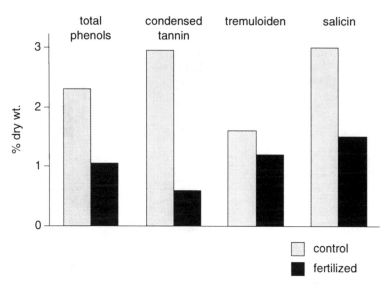

Figure 2.20. Effects of nitrogen fertilization on levels of two simple phenolics (salicin and tremuloiden), condensed tannins, and total phenols in foliage of quaking aspen, *Populus tremuloides*. Fertilization reduces the amounts of all the phenols (after Bryant et al., 1987).

Furthermore, the same varieties, grown in Britain, where light levels are relatively low, especially in greenhouses, had lower levels of phenolics. Other important sources of variation in phenolics are the levels of soil nutrients, especially nitrogen. Poor soils tend to cause an increase in the levels of these compounds, and this has been recorded especially for phenols in many species of trees. For example, Bryant and his co-workers showed reduced levels of tannin and two phenolic glycosides, salicin and tremuloiden, in quaking aspen, as a result of fertilization (Fig. 2.20). High levels of CO_2 also increase the carbon:nitrogen ratio in plants and this is reflected in the changes seen in these "carbon-based" secondary metabolites.

Damage caused by insects or pathogens can induce the production of phenols which may be measurable after hours. Additional compounds seem to be produced over a longer period and even after a year or more the induced changes may be observed in certain tree species (Fig. 2.21).

Genetic differences in phenolic content of plants have been demonstrated in

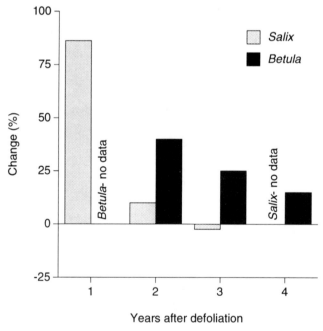

Figure 2.21. Increases in levels of phenols in foliage of birch *(Betula)* and willow *(Salix)* trees, 1 to 4 years after defoliation in Finland. All trees were completely defoliated at the start. In the following year phenol levels were 75% higher than the controls in willow, and back to normal the year after. In birch, levels were still significantly higher than controls after four years (after Tuomi et al., 1991).

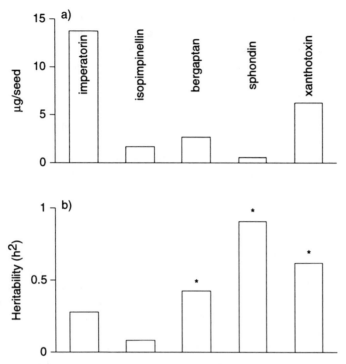

Figure 2.22. Genetic variation in levels of furanocoumarins in wild parsnip, *Pastinaca sativa*. **a)** mean concentrations of five furanocoumarins in plants from a wild population. **b)** seeds were collected from individual plants and grown to maturity in a greenhouse. The amounts of furanocoumarins in seeds of these plants, and those of their offspring, were compared by regression analysis, and the heritability estimated as two times the regression slope. Three of the furanocoumarins had significant heritabilities as shown with asterisks (after Zangerl et al., 1989).

many species. For example, different varieties of crop plants have been found to vary in phenolic acid profiles or levels. An elegant demonstration of genetic differences in furanocoumarins of wild parsnip, *Pastinaca sativa*, illustrates the potential for short term evolutionary change (Fig. 2.22). Amounts of the different compounds were measured in half sibs from a wild population, and from their offspring. Heritability of the quantity of some furanocoumarins was significant, especially for those in the seeds.

2.4.3 Terpenoids

These are ubiquitous and form the largest and most diverse class of organic compounds found in plants, with over 15,000 characterized so far. They share a common biosynthetic origin, with the fusion of five-carbon unsaturated hydro-carbon units, called isoprene units. They range from essential oil components to nonvolatile triterpenoids and sterols, and carotenoid pigments. They are mostly lipid soluble and located in the cytoplasm or in special glands.

2.4.3.1 Mono-, sesqui- and diterpenoids

Monoterpenes (10 carbon atoms) are extremely common odoriferous compounds like citral, menthol and geraniol. They are often also important floral odors at low concentrations. Fig. 2.23 gives some of the structures. There are about 1,000 different known monoterpenes, and an even greater number of compounds related to them. They occur in 57 plant families. They are most widely recognized as constituents of conifers, Lamiaceae (=Labiatae) such as mints, Asteraceae (=Compositae) such as *Ambrosia*, and Rutaceae such as *Citrus*. Some compounds are regularly found together in leaf oils: for example pinene and limonene. Isomerism is common and simple optical isomers may have totally different odors. For example, l-carvone gives spearmint its characteristic odor, while d-carvone gives caraway its particular odor. Monoterpenes are particularly significant for distance attraction of insect herbivores because of their high volatility. In addition, some of them are major constituents of floral odors, like geraniol in roses and geraniums. Another related group, the iridoids, have an additional lactone (5-membered ring) and occur as glycosides.

Sesquiterpenes (15 carbon atoms) are also often volatile. Examples are caryophyllene in conifers and gossypol in cotton. The sesquiterpene lactones are widespread

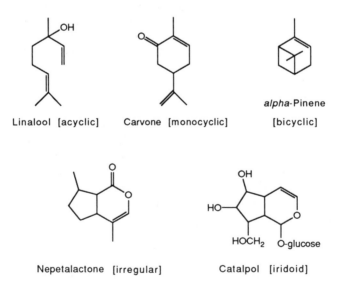

Linalool [acyclic]   Carvone [monocyclic]   *alpha*-Pinene [bicyclic]

Nepetalactone [irregular]   Catalpol [iridoid]

Figure 2.23.   Structures of some monoterpenoids.

in the Asteraceae (=Compositae) and are known to be deterrent to many insects as well as bitter to humans (see Fig. 2.24 for some of the structures). There are at least 4,500 different sesquiterpenes, including 3,500 sesquiterpene lactones.

Diterpenes (20 carbon atoms) often occur in resins or latex. Abietic acid (Fig. 2.24), for example, is widespread in gymnosperm resins, as are many of the other 2,000 types. The clerodanes occur in several families of herbaceous plants and include the compound clerodendrin from *Clerodendrum*. The drimanes form another big group and include the compounds warburganol from *Warburgia* and polygodial from *Polygonum*. These three compounds are feeding deterrents to many different insect species.

2.4.3.2 Triterpenoids

The triterpenoids (30 carbon atoms) include an enormous variety of complex structures based on the four-ring triterpene structure but with a great variety of

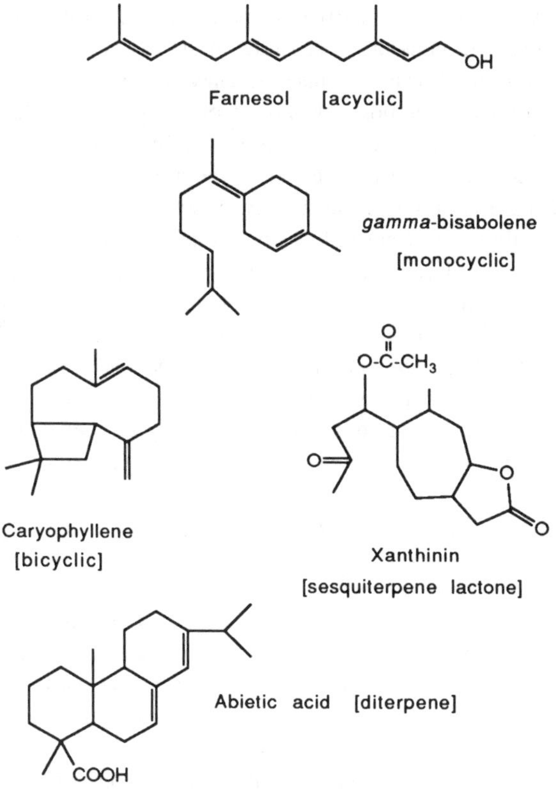

Farnesol   [acyclic]

gamma-bisabolene

[monocyclic]

Caryophyllene
[bicyclic]

Xanthinin
[sesquiterpene lactone]

Abietic acid [diterpene]

Figure 2.24. Structures of some sesquiterpenes and a diterpene.

additions (Fig. 2.25). They include true triterpenes, such as amyrin, steroids, steroidal glycosides (saponins and cardenolides) and steroidal alkaloids. This class of compounds includes at least 3,000 structures. With a recent focus on this group, as a result of improved analytical techniques, many new ones are being discovered each year.

Some of the true triterpenes occur in waxes. For example, β-amyrin and ursolic acid are common wax components. The cucurbitacins in members of the family Cucurbitaceae and other plants are well known. Perhaps the best known insect deterrent is azadirachtin, a limonoid triterpene in the neem tree, *Azadirachta indica*. It deters many species of insects, and, in particular, the desert locust. It may be that the triterpenoids will turn out to be of major importance in plant-insect interactions just as alkaloids appear to be of major importance in their effects on mammals.

Phytecdysteroids, compounds that are related to insect molting hormone, have been found at 2 or 3% dry weight in roots and in bark, but, in the approximately 100 species where they have been found in leaves, the levels are usually less than 0.01% dry weight, and usually have no behavioral or physiological significance for insects. Saponins are glycosides of sterols and triterpenes and some have been found to be deterrent to insects. They are surface-active agents (soaplike) and may influence the post-ingestive properties of other plant compounds. For example, they may reduce the ability of tannins to bind with proteins, but, on the other hand, they bind to free sterols themselves.

Cardenolides are common in the family Asclepiadaceae, but occur also in a dozen other families. They are potent vertebrate poisons. Cardenolides usually occur as glycosides, often with unusual sugar moieties. They are deterrent to many insects, but are most famous among insects for their sequestration as defensive compounds by specialists such as the monarch butterfly, *Danaus plexippus*, the milkweed bug, *Oncopeltus fasciatus*, and the oleander aphid, *Aphis nerii*. In fact, there is a large fauna associated with milkweeds and the majority of species sequester cardenolides and are brightly colored. The accumulated compounds in the body are potent poisons for many vertebrate predators.

2.4.3.3 Variation in levels of terpenoids

Terpenoid concentrations are often highest in reproductive structures. In trees, however, the highest levels may be in the trunk. In addition, since terpenoid-containing plants usually have a suite of related chemical structures, it is common for different organs to have different proportions of the individual chemicals. As with many other secondary metabolites, young leaves have higher levels than mature leaves.

Increases in light intensity usually result in elevated terpenoid concentrations, just as with phenolics. Some authors postulate that the additional photosynthate favors production of non-nitrogenous secondary metabolites. The precise physiological controls are not yet understood, but large increases do occur both in the

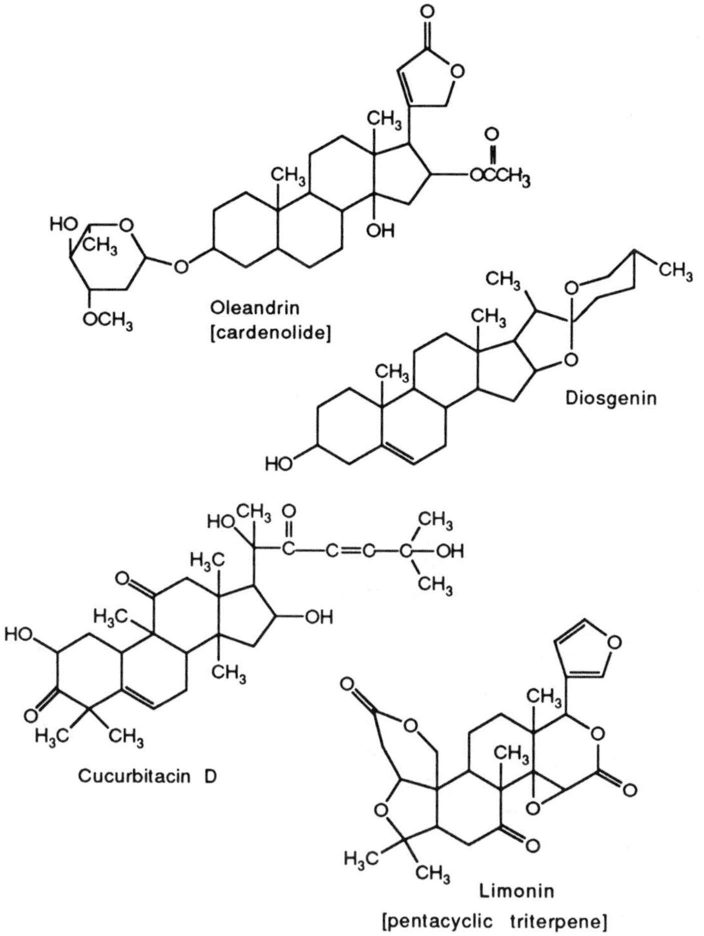

Figure 2.25. Structures of selected triterpenes.

short and long term. It is well known, for example, that yields of essential oils in mints and lavenders are much higher when plants are harvested at the end of long, hot, sunny days, than after cool, cloudy days. They also accumulate to high levels under drought stress. Fertilization with nitrogen, phosphorus or potassium generally leads to lower terpenoid concentrations. For example, experiments with camphorweed grown under different levels of nutrient nitrogen showed threefold differences in volatile terpene yield (Table 2.8).

Genetic variation in concentrations and types of terpenoids occurs in many plants. In particular, many genotypes of mint species are known with different quantities and profiles of monoterpenes. Variation in quantity and quality of monoterpenes in conifers is under genetic control, although environmental effects

Table 2.8. *Influence of nitrogen availability on leaf terpene content in camphorweed,*
Heterotheca subaxillaris, *after different lengths of time in the treatments (after*
Mihaliak and Lincoln, 1985).

Nitrate concentration in growth medium	Time in weeks	Leaf nitrogen (mg/g)	Leaf terpene (mg/g)
5×10^{-4}M	1	20	7.5
	3	18	5.8
	5	15	4.2
	7	12	1.8
1.5×10^{-2}M	1	44	4.0
	3	41	2.9
	5	32	2.8
	7	30	2.6

also contribute to phenotypic differences, suggesting the potential for strong
genotype-by-environment interaction effects on monoterpene composition.

Considerable amounts of work have been carried out on variation in the
content of sesquiterpene lactones in *Ambrosia* spp. with the aim of understanding
polymorphisms, their origin, significance and adaptive value. This is a herbaceous
genus in the family Asteraceae, common in North America. Two species, *A.cu-
manensis* and *A.psilostachya* have been studied in detail around the Gulf of
Mexico. Sixty-two populations of *A.psilostachya* were analyzed and it was found
that, although over a dozen sesquiterpene lactones characterize the species as a
whole, any single population contains only one or two as major components.
Differences are genetic, and there is often overlap in geographic distribution of
the different types. Variation also occurs in *A.cumanensis*. In Fig. 2.26 the
distribution of some chemotypes is illustrated.

Among triterpenoids, the inheritance of cucurbitacins in various species of
Cucurbitaceae has been established. Several cultivated cucurbits have been studied in detail because cucurbitacins are so bitter and toxic to humans, and in
cucumber, for example, the absence of cucurbitacins is known to be controlled
by a recessive gene. The levels of cardenolides in *Asclepias syriaca* vary greatly.
Some plants contain none while others reach levels of 2mg/g dry weight (Fig.
2.27). Monarch butterflies prefer to oviposit on the plants with intermediate
concentrations.

2.4.4 Organic acids, lipids and related compounds

A number of long chain fatty acids characterize certain plant groups. For example
erucic acid (C22) occurs in the Brassicaceae (= Cruciferae) and Tropaeolaceae.
On the other hand oleic acid (C16) is widespread and comprises 80% of the fatty
acid content of olive oil (Fig. 2.28).

Polyacetylenes are hydrocarbons which all have one or more acetylenic groups

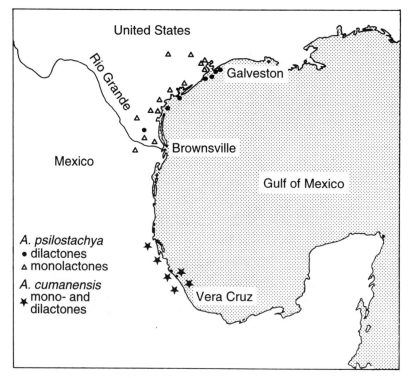

Figure 2.26. Occurrence of mono- and dilactone sesquiterpenes in two species of *Ambrosia*. Local populations of *A. psilostachya* have either one or the other type. Plants of *A.cumanensis*, from Mexico, have both types (after Smith, 1976).

in their structures (Fig. 2.29). They often have additional functional groups. There are about 650 known structures. They occur regularly in five families: Asteraceae, Campanulaceae, Apiaceae, Pittosporaceae and Araliaceae. Some are known to be toxic to microorganisms and animals, but little is known about their effects on insect behavior, except that in some cases polyacetylenes provide specific phagostimulatory information (see Chapter 4).

2.4.5 Sulfur-containing compounds

In addition to the sulfur-containing amino acids, cysteine and methionine, there are other sulfur-containing organic compounds with restricted distribution that are known to influence insect host selection. The most important are the glucosinolates of the Brassicaceae (= Cruciferae) and the organic disulfides of *Allium*. The glucosinolates are the *in vivo* precursors of the mustard oils, or isothiocyanates, and yield these acrid volatiles after enzymic hydrolysis. Whenever insects chew on such plants, myrosinase acts on the glucosinolates to release isothiocya-

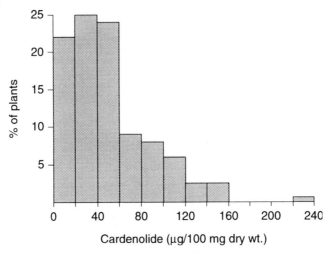

Figure 2.27. Variation in levels of cardenolides in common milk-weed. Most plants have relatively low levels and some have none at all, but, in a few, the cardenolide level is very high. Variation is due to a combination of genetic and environmental factors (after Malcolm et al., 1989).

Saturated acids

C20   Arachidic $CH_3(CH_2)_{18}COOH$

Unsaturated acids

C18   Linoleic $CH_3(CH_2)_4CH=CHCH_2CH=CH(CH_2)_7COOH$

C18   Linolenic $CH_3CH_2CH=CHCH_2CH=CHCH_2CH=CH(CH_2)_7COOH$

Unusual acids

C22   Erucic $CH_3(CH_2)_7CH=CH(CH_2)_{11}COOH$

C19   Sterculic $CH_3(CH_2)_7C{=}C(CH_2)_7COOH$
$$\underset{CH_2}{\diagdown\diagup}$$

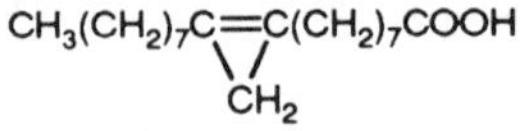

Figure 2.28. Structures of fatty acids.

ACETYLENES

Straight chain
hydrocarbon

$CH_3CH=CH(C≡C)_2CH=CH=CH=CH_2$

Alcohols

$CH_3CH=CH(OH)C≡CC≡CCH_2CH=CH(CH_2)_6CH_3$

Falcarinol

Ketone

$CH_2CHCO(C≡C)_2CH_2CH=CH(CH_2)_6CH_3$

Falcarinone

Fatty acid

$CH_3(CH_2)_7 C≡C(CH_2)_7COOH$

Stearolic acid

Figure 2.29. Structures of acetylenes.

nates (Fig. 2.30). There are about seventy known, and the sulfur-linked sugar is always glucose. Apart from Brassicaceae they occur in a number of other families, especially in the order Capparales. Elsewhere, they occur sporadically.

Perhaps because of the long history of *Brassica* crops, the chemistry of glucosinolates has been well studied. Distinctive arrays of similar chemicals occur among closely related taxa, but the glucosinolate profiles are species-specific.

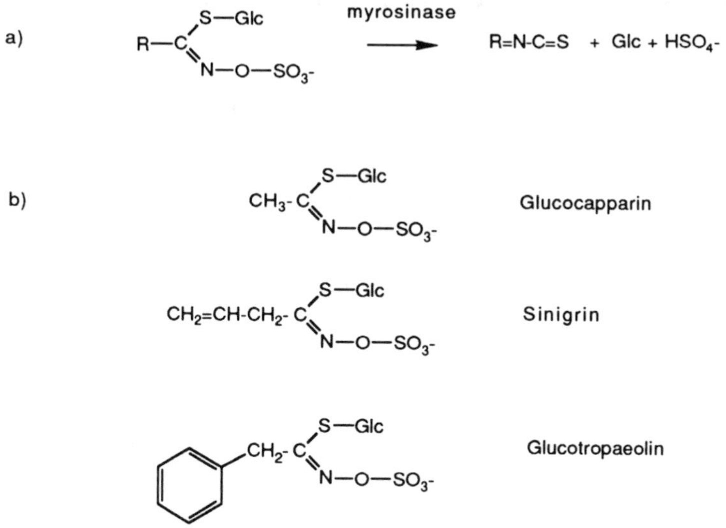

a)

$$R-C \begin{array}{c} S—Glc \\ \\ N—O—SO_3^- \end{array} \quad \xrightarrow{\text{myrosinase}} \quad R=N-C=S \; + \; Glc + HSO_4^-$$

b)

$$CH_3-C \begin{array}{c} S—Glc \\ \\ N—O—SO_3^- \end{array} \qquad \text{Glucocapparin}$$

$$CH_2=CH-CH_2-C \begin{array}{c} S—Glc \\ \\ N—O—SO_3^- \end{array} \qquad \text{Sinigrin}$$

$$CH_2-C \begin{array}{c} S—Glc \\ \\ N—O—SO_3^- \end{array} \qquad \text{Glucotropaeolin}$$

Figure 2.30. **a)** Release of thiocyanate (R=N-C=S) by the action of myrosinase on a glucosinolate. Free glucose (Glc) is also produced. **b)** Three different glucosinolates.

In addition, the breakdown products of a given glycoside vary depending on co-factors and pH.

The sulfides (Fig. 2.31) are simple compounds such as allyldisulfide character-istic of garlic, and propyldisulfide, characteristic of onion. These odors are released from bound forms in the plant, especially as a result of damage. The sulfides may be oxidized to sulfoxides and sulfones. Thiophenes (Fig. 2.31) occur in a number of species of Asteraceae.

Considerable work has been carried out on quantitative variation in concentra-tion of glucosinolates. In general, seedlings have higher levels than older plants; reproductive tissue has higher levels than foliage; spring foliage has higher levels than late summer foliage; soil sulfate levels are positively correlated with plant glucosinolate levels. The Rocky Mountain bittercress, *Cardamine cordifolia*, is one of the plants that has been studied in which several factors have been found to determine levels of glucosinolates (Fig. 2.32). In addition to the total levels of glucosinolates seen, there was variation in proportions of the different chemicals. Many crop plants are in the families Brassicaceae and Alliaceae, and, as a result, genetic variants with respect to sulfur compounds in both families are well known. For example, heritabilities of 35 to 80% have been found for individual glucosinolates of rape, *Brassica napus*.

2.5 Conclusions

No two plants are chemically identical. Nutritionally, the plant is always in a state of flux. There are hundreds of thousands of different secondary metabolites.

Sulphides

CH_3-S-S-CH_3 Dimethyldisulphide

CH_2=CH-CH_2-S-S-CH_2-CH=CH_2 Diallyldisulphide

CH_2=CH-CH_2-S-S-CH_3 Methylallyldisulphide

CH_2=CH-S-CH=CH_2 Diallylsulphide

Thiophenes

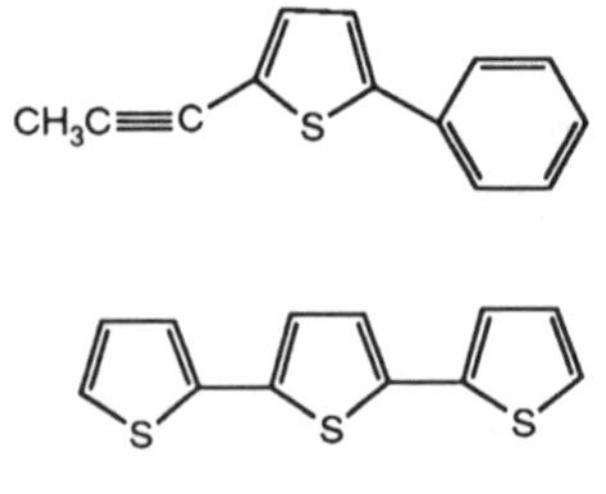

Figure 2.31.   Structures of sulfur compounds in plants; sulfides and thiophenes.

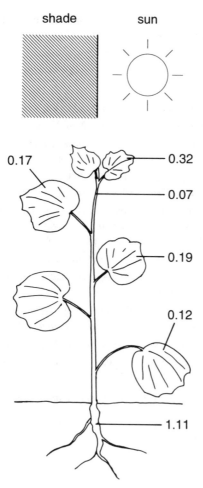

Figure 2.32. Diagram illustrating variation in the total levels of glucosinolates in Rocky Mountain bittercress, *Cardamine cordifolia*, in relation to position on the plant and the degree of insolation (after Louda and Rodman, 1983).

Even within a plant species the compounds present vary both qualitatively and quantitatively. In addition, the environmental factors influencing plant chemistry are constantly varying and there are differential effects of environmental variables on different chemical types. Intraspecific genetic variation for individual chemicals has been found wherever it has been looked for.

Plants provide a diet for insects which is extremely variable, and much more so than other types of diets. A challenge for insect herbivores selecting their host plants is being able to distinguish between the relevant chemistry and the noise. This must be done by the sensory system, which is the subject of Chapter 3.

Further reading

Barbosa, P., Krischik, V.A. and Jones C.G. (eds.) 1991. *Microbial Mediation of Plant-Herbivore Interactions*. Wiley Interscience, New York.

Denno, R.F. and McClure, M.S. (eds.) 1984. *Variable Plants and Herbivores in Natural and Managed Systems*. Academic Press, New York.

Fritz, R.S. and Simms, E.L. (eds.) 1992. *Plant Resistance to Herbivores and Pathogens: Ecology, Evolution and Genetics*. Chicago University Press.

Gulman, S.L. and Mooney, H.A. 1986. Costs of defense and their effects on plant productivity. In Givnish, T. (ed.) *On the Economy of Plant Form and Function*. Cambridge University Press, Cambridge, pp. 681–698.

Harborne, J. 1988. *Introduction to Ecological Biochemistry*. Academic Press, London.

Pearcy, R.W., Ehleringer, R., Mooney,H.A. and Rundel,P.A. (eds.) 1990. *Plant Physiological Ecology*. Chapman & Hall, New York.

Rosenthal, G.A. and Berenbaum, M.R. (eds.) 1992. *Herbivores. Their Interactions with Secondary Metabolites, vol. 1. The Chemical Participants*. Academic Press, New York.

Smith, P.M. 1976. *The Chemotaxonomy of Plants*. Arnold, London.

Tallamy, D.W. and Raupp, M.J. (eds.) 1991. *Phytochemical Induction by Herbivores*. Wiley, New York.

Vickery, M.L. and Vickery, B. 1981. *Secondary Plant Metabolism*. MacMillan, London.

References (* indicates review)

Volatiles

Gibbs, R.D. 1974. *The Chemotaxonomy of Flowering Plants*. McGill-Queen's University Press, Montreal.

*Metcalf, R.L. 1987. Plant volatiles as insect attractants. CRC Rev.Plant Science 5: 251–301.

Visser, J.H. and Avé, D.A. 1978. General green leaf volatiles in the olfactory orientation of the Colorado potato beetle, *Leptinotarsa decemlineata*. Entomologia Exp.Appl. 24: 738–749.

Surfaces, surface waxes and other surface compounds

*Baker, E.A. 1982. Chemistry and morphology of plant epicuticular waxes. *In* Cutler,D.F., Alvin, K.L. and Price, C.E. (eds.) *The Plant Cuticle*. Academic Press, London, pp. 139–165.

*Brown, V.K. and Lawton, J.H. 1991. Herbivory and the evolution of leaf size and shape. Phil.Trans.R.Soc.Lond.B. 333: 265–272.

*Chapman, R.F. 1977. The role of the leaf surface in food selection by acridids and other insects. Coll.Int. CNRS 265: 133–150.

Eigenbrode, S.D., Espelie, K.E. and Shelton, A.M. 1991. Behavior of neonate diamond-back moth larvae (*Plutella xylostella*) on leaves and on extracted leaf waxes of resistant and susceptible cabbages. J.Chem.Ecol. 17: 1691–1704.

Juniper, B.E. and Jeffree, C.E. 1983. *Plant Surfaces*. Edward Arnold, London.

Juniper, B. and Southwood, R. (eds.) 1986. *Insects and the Plant Surface*. Edward Arnold, London.

Martin, J.T. and Juniper, B.E. 1970. *The Cuticles of Plants*. Arnold, London.

Preece, T.F. and Dickinson, S.H. (eds.) 1971. *Ecology of Leaf Surface Micro-Organisms*. Academic Press, London.

Internal Components: nutrients

Anderson, P.C., Brodbeck, B.V. and Mizell, R.F. 1992. Feeding by the leafhopper, *Homalodisca coagulata*, in relation to xylem fluid chemistry and tension. J.Insect Physiol. 38: 611–623.

*Bernays, E.A. 1992. Plant sterols and host-plant affiliation of herbivores. *In* Bernays, E.A. (ed.) *Insect Plant Interactions*, vol. 4. CRC Press, Boca Raton, pp. 45–58.

Crawley, M.J. 1983. *Herbivory*. Blackwell, Oxford.

Cull, D.C. and van Emden, H.F. 1977. The effect on *Aphis fabae* of diel changes in their food quality. Physiol.Entomol. 2: 109–115.

Faeth, S., Mopper, S.H. and Simberloff, D. 1981. Abundances and diversity of leaf-mining insects on three oak species: effects of host plant phenology and nitrogen content of leaves. Oikos 37: 238–251.

Feeny, P. 1970. Seasonal changes in oak leaf tannins and nutrients as a cause of spring feeding by winter moth caterpillars. Ecology 51: 565–581.

Mattson, W.A. 1980. Herbivory in relation to plant nitrogen content. A.Rev.Ecol. 11:19–38.

*McNeill, S. and Prestige, R.A. 1982. Plant nutritional strategies and insect herbivore community dynamics. In Visser, J.H. and Minks A.K. (eds.) *Insect-Plant Relationships*. Pudoc, Wageningen, pp. 225–235.

*Pate, J.S. 1983. Patterns of nitrogen metabolism in higher plants and their ecological significance. *In* Lee, J.A., McNeill, S. and Rorison, I.H. (eds.) *Nitrogen as an Ecological Factor*. Blackwell, Oxford, pp. 225–255.

Prestige, R.A. 1982. Instar duration, adult consumption, oviposition and nitrogen utilization efficiencies of leafhoppers feeding on different quality food (Auchenorrhyncha: Homoptera). Ecol.Entomol. 7:91–101.

Shaw, G.G. and Little, C.H.A. 1972. Effect of high urea fertilization of balsam fir trees on spruce budworm. *In* Rodriguez, J.G. (ed.) *Insect and Mite Nutrition*. North Holland Publishing Company, Amsterdam, pp. 589–598.

*Slansky, F. and Scriber, J.M. 1984. Food consumption and utilization. *In* Kerkut, G.A. and Gilbert, L.I. (eds.) *Comprehensive Insect Physiology, Biochemistry and Pharmacology* vol. 4. Pergamon Press, Oxford, pp. 87–163.

Soldaat, L.L. 1991. Nutritional Ecology of *Tyria jacobaeae* L. PhD thesis, University of Leiden.

Waite, R. and Boyd, J. 1953. The water-soluble carbohydrates of grasses. J.Sci.Food Agric. 4: 197–205.

Secondary metabolites: general

*Bryant, J.P., Tuomi, J. and Niemala, P. 1988. Environmental constraint of constitutive and long term inducible defences in woody plants. *In* Spenser, K.C. (ed.) *Chemical Mediation of Coevolution*. Academic Press, New York, pp. 367–390.

Cutler, D.F. 1978. *Applied Plant Anatomy*. Longmans, London.

Dussourd, D.E. and Denno, R.F. 1991. Deactivation of plant defense: correspondence between insect behavior and secretory canal architecture. Ecology 72: 1383–1396.

*Gregory, P., Avé, D.A., Bouthyette, P.J. and Tingey, W.M. 1986. Insect-defensive chemistry of potato glandular trichomes. *In* Juniper, B. and Southwood, T.R.E. (eds.) *Insects and the Plant Surface*. Arnold, London, pp. 173–184.

Harborne, J. 1986. *Phytochemical Methods. A Guide to Modern Techniques of Plant Analysis*. Chapman and Hall, London.

Ihlenfeldt, H.D. and Hartmann, H.E.K. 1982. Leaf surfaces in Mesembryanthemaceae. *In* Cutler, D.F., Alvin, K.L. and Price, C.E. (eds.) *The Plant Cuticle*. Academic Press, New York, pp. 397–424.

Parrot, W.L., Jenkins, J.N.and McCarty, J.C. 1983. Feeding behavior of first-stage tobacco budworm (Lepidoptera: Noctuidae) on three cotton cultivars. Ann.Entomol.Soc.Am. 76: 167–170.

Rodriguez, E., Healy, P.L. and Mehta, E. (eds.) 1983. *Biology and Chemistry of Plant Trichomes*. Plenum Press, New York.

Rosenthal, G. and Berenbaum, M. (eds.) 1992. *Herbivores: Their Interactions with Secondary Plant Metabolites* vol. 1. Academic Press, New York.

*Schnepf, E. 1976. Morphology and cytology of storage spaces. *In* Luckner, M., Mothes, K. and Nover, L. (eds.) *Secondary Metabolism and Coevolution*. Barth, Leipzig, pp. 23–44.

*Zangerl, A.R. and Bazzaz, F.A. 1991. Theory and pattern in plant defense allocation. *In* Fritz, R.S. and Simms, E.L. (eds.) *Plant Resistance to Herbivores and Pathogens*. Chicago University Press, pp. 363–391.

Secondary metabolites: nitrogen containing

Baldwin, I.T. 1989. Mechanism of damage-induced alkaloid production in wild tobacco. J.Chem Ecol. 15: 1661–1680.

Bernays, E.A. and Chapman, R.F. 1976. Antifeedant properties of seedling grasses. *In* Jermy, T. (ed.) *The Host Plant in Relation to Insect Behaviour and Reproduction.* Akademiai Kiado, Budapest, pp. 41–46.

*Bernays, E.A. 1983. Nitrogen in defence against insects. *In* Lee, J.A., McNeill, S. and Rorison, I.H. (eds.) *Nitrogen as an Ecological Factor.* Blackwell, Oxford, pp. 321–344.

*Clay, K. 1992. Fungal endophytes, grasses and herbivores. *In* Barbosa, P., Krischik, V. and Jones, C.G. *Microbial Mediation of Plant-Herbivore Interactions.* Wiley, New York, pp. 199–226.

Conn, E.E. 1979. Cyanide and cyanogenic glycosides. *In* Rosenthal, G.A. and Janzen, D.H. (eds.) *Herbivores: Their Interaction with Secondary Plant Metabolites.* Academic Press, New York, pp. 378–412.

Cooper Driver, G., Finch, S., Swain, T. and Bernays, E.A. 1977. Seasonal variation in secondary plant compounds in relation to the palatability of *Pteridium aquilinum.* Biochem.Syst.Ecol. 5: 177–183.

*Dahlman, D.L., Eichenseer, H. and Siegel, M.R. 1991. Chemical perspectives of endophyte-grass interactions and their implications to insect herbivory. *In* Barbosa, P., Krischik, V.A. and Jones, C.G. (eds.) *Microbial Mediation of Plant-Herbivore Interactions.* Wiley, New York, pp. 227–252.

*Hartmann, T. 1992. Alkaloids. *In* Rosenthal, G.A. and Berenbaum, M.R. (eds.) *Herbivores: Their Interaction with Secondary Plant Metabolites,* vol. 1. Academic Press, New York, pp. 79–122.

Vrieling, K. 1990. Costs and benefits of alkaloids of *Senecio jacobaea* L. PhD thesis, University of Leiden.

*Wink, M. 1992. The role of quinolizidine alkaloids in plant-insect interactions. *In* Bernays, E.A. (ed.) *Insect Plant Interactions* vol. 4. CRC Press, Boca Raton, pp. 131–165.

Secondary metabolites: phenolics

*Berenbaum, M.R. 1992. Coumarins. *In* Rosenthal, G.A. and Berenbaum, M.R. (eds.) *Herbivores: Their Interactions with Secondary Metabolites,* vol. 1. Academic Press, New York, pp. 221–250.

Berenbaum, M.R., Zangerl, A.R. and Nitao, J.K. 1986. Constraints on chemical coevolution: wild parsnips and the parsnip webworm. Evolution 40: 1215–1228.

*Bernays, E.A., Cooper Driver, G. and Bilgenor, M. 1989. Herbivores and plant tannins. Adv.Ecol.Res. 19: 263–302.

Bryant, J.P., Clausen, T.P., Reichardt, P.B., McCarthy, M.C. and Werner, R.A. 1987.

Effect of nitrogen fertilization upon the secondary chemistry and nutritional value of quaking aspen leaves for the large aspen tortrix. Oecologia 73: 513–517.

*Hagerman, A.E. and Butler, L.G. 1992. Tannins and lignins. *In* Rosenthal, G.A. and Berenbaum, M.R. (eds.) *Herbivores: Their Interactions with Secondary Metabolites,* vol. 1. Academic Press, New York, pp. 355–387.

*Harborne, J.B. 1988. Flavonoids. *In* Goodwin, T.W. (ed.) *Plant Pigments.* Academic Press, London, pp. 299–243.

Swain, T. 1974. Reptile-angiosperm coevolution. *In* Luckner, M., Mothes, K. and Nover, L. (eds.) *Secondary Metabolism and Coevolution.* Barth, Leipzig, pp. 551–561.

*Tuomi, J., Fagerstrom, T. and Niemala, P. 1991. Carbon allocation, phenotypic plasticity, and induced defenses. *In* Tallamy, D.W. and Raupp, M.J. (eds.) *Phytochemical Induction by Herbivores.* Wiley Interscience, New York, pp. 85–104.

*Waterman, P.G. and Mole, S. 1989. Extrinsic factors influencing production of secondary metabolites in plants. *In* Bernays, E.A. (ed.) *Insect-Plant Interactions* vol. 1. CRC Press, Boca Raton, pp. 107–134.

Zangerl, A.R., Berenbaum, M.R. and Levine, E. 1989. Genetic control of seed chemistry and morphology in wild parsnip (*Pastinaca sativa*). J.Heredity 80: 404–407.

Secondary metabolites: terpenoids

*Gershenzon, J. and Croteau, R. 1992. Terpenoids. *In* Rosenthal, G.A. and Berenbaum, M.R. (eds.) *Herbivores: Their Interactions with Secondary Metabolites*, vol. 1. Academic Press, New York, pp. 165–220.

Harborne, J.B. (ed.) 1991. *Ecological Chemistry and Biochemistry of Plant Terpenoids.* University Press, Oxford.

Malcolm, S.B., Cockrell, B.J. and Brower, L.P. 1989. Cardenolide fingerprints of monarch butterflies reared on common milkweed *Asclepias syriaca*. J.Chem.Ecol.15: 819–853.

Mihaliak, C.A. and Lincoln, D.E. 1985. Growth pattern and carbon allocation to volatile leaf terpenes under nitrogen limiting conditions in *Heterotheca subaxillaris* (Asteraceae). Oecologia 66: 423–426.

Secondary metabolites: lipids and sulfur-containing

*Seigler, D.S. 1979. Toxic seed lipids. *In* Rosenthal, G.A. and Janzen, D.H. (eds.) *Herbivores: Their Interaction with Secondary Plant Metabolites.* Academic Press, New York, pp. 449–470.

*Etten, C.H.Van and Tookey, H.L. 1979. Chemistry and biological effects of glucosinolates. *In* Rosenthal, G.A. and Janzen, D.H. (eds.) *Herbivores: Their Interaction with Secondary Plant Metabolites.* Academic Press, New York, pp. 471–500.

Louda, S.M. and Rodman, J.E. 1983. Ecological patterns in the glucosinolate content of a native mustard, *Cardamine cordifolia*, in the Rocky Mountains. J.Chem.Ecol. 9: 397–421.

3

Sensory Systems

In selecting its host plant an insect may use a variety of senses: smell and taste, vision and touch. Probably all of these are used by most insects at some stage in the host-selection process, although, at any one stage, one sense may predominate. In this chapter we describe the physiological bases of sensory perception, concentrating on those aspects that are particularly important in host-plant selection.

3.1 Introduction

The basic unit of any sensory system is the sense cell, or neuron. In insects, the cell bodies of the sense cells are usually in the epidermis immediately beneath the cuticle. From the cell body a short, fingerlike process, the dendrite, extends towards the outside, and a long slender axon runs without any synapses into the central nervous system. Within the central nervous system the axon branches extensively; it is said to arborize. (The visual system does not conform to this pattern (see below).) The dendrite is modified to interact with the environment so that an appropriate stimulus produces a change in the electrical potential across the cell membrane. This change in energy from one form, e.g., light or mechanical, to another, electrical, is called transduction. The potential created by stimulation of the sense cell is called the receptor potential (Fig. 3.1).

The insect, like any other animal, must differentiate between stimulus quality and quantity. Detection of the stimulus quality involves not only discriminating between, say, light and a chemical stimulus, but also discriminating between different wavelengths of light or types of chemical. This is achieved by having cells that are sensitive to the different modalities and which are sensitive to different ranges within a modality. Thus, a visual cell may be sensitive only to ultraviolet light or to a specific range of wavelengths in the human visible

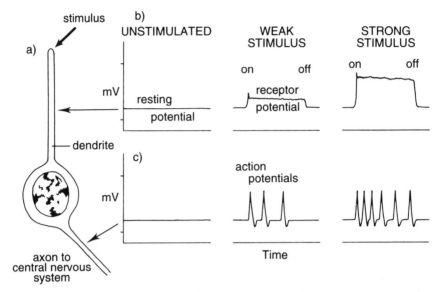

Figure 3.1. Basic anatomy and physiology of a receptor system. **a)** The sensory neuron is the basic unit in any receptor system. In insects, the sensory neurons are in the epidermis. An appropriate stimulus reaching the dendrite causes a change in the resting potential of its membrane, the receptor potential. **b)** The receptor potential may be large or small, depending on the magnitude of the stimulus. The receptor potential spreads to the cell body and, somewhere near the origin of the axon, action potentials are produced. **c)** Action potentials only last for 2-3 ms. The number produced is proportional to the magnitude of the receptor potential, but their size is constant. The number and size of the action potentials reaching the end of the axon in the central nervous system is the same as that leaving the cell body. Action potentials are also sometimes called spikes or impulses.

spectrum. A chemosensitive cell may be sensitive primarily to hexose sugars, or inorganic salts, or to some other class of chemicals. Sense cells may respond to a broad spectrum of stimuli within a modality, or be sharply tuned, only responding to specific stimuli (see below).

Stimulus quantity is correlated with the magnitude of the receptor potential. A strong stimulus causes a large receptor potential, while a weak stimulus results in a small one. Because the size of the potential varies in relation to stimulus intensity, it is said to be graded.

The receptor potential spreads relatively slowly over the cell membrane and a further process is necessary to transmit information rapidly to the central nervous system. This process is the production of action potentials. An action potential is a very brief change in the membrane potential lasting only 2–3ms. It moves rapidly along the axon without changing; unlike the receptor potential it is not graded so that the information arriving at the end of the axon in the central nervous system is precisely the same as that which leaves the body of the sense cell. Information about the quantity of a stimulus is given by the number of action potentials produced per unit of time, which is proportional to the size of the receptor potential (Fig. 3.1).

In most cases, a sense cell only continues to produce action potentials for a very brief period, often less than one second, even though it continues to be stimulated. The cell is said to become adapted and requires a period without stimulation before it will respond again with the same level of activity.

3.2 The sense of taste

The sense of taste in insects is more appropriately referred to as "contact chemoreception" because it differs in some respects from the common perception of taste in vertebrates. In vertebrates, taste receptors occur in the oral cavity, especially on the tongue. In insects, they occur outside the mouth, and are often present on the feet and other parts of the body. Further, whereas in our sense of taste the chemical stimuli are usually present in solution in saliva, insects can perceive chemicals on dry surfaces, such as the surface of a leaf. Contact chemoreceptors have the form of hairs or small cones of cuticle with a single pore at the tip. From the pore, a tube of cuticle-like material extends down inside the hair, and within this tube, the dendrite sheath, are the dendrites of the sensory cells. Commonly there are four or six sense cells associated with a single hair and their dendrites run through the dendrite sheath, ending just inside the pore (Fig. 3.2). The endings of the dendrites are embedded in a mucopolysaccharide through which any stimulating molecule must pass in order to stimulate the dendrite. The nature of this mucopolysaccharide is thus critically important, but we know almost nothing about it.

The whole structure of hair and its socket, sense cells and supporting cells which surround the sense cells is called a sensillum.

Each of the sense cells within a sensillum is sensitive to a different range of chemicals. Commonly, one cell responds to sugars, one to inorganic salts, one to behaviorally deterrent compounds, and one to water or amino acids. This does not necessarily imply that the cells respond only to these groups of chemicals, but that they are more sensitive to them and so are most likely to be actively stimulated by the appropriate compound.

In no case do we have certain knowledge of the mechanism by which a compound elicits a response from a particular cell, and most work on this topic has been on flies rather than on phytophagous insects. It is most likely that specific acceptor proteins occur in the dendrite membrane. These proteins associate with potential stimulating molecules, having specific molecular configurations, so that only molecules with these configurations are capable of initiating a receptor potential.

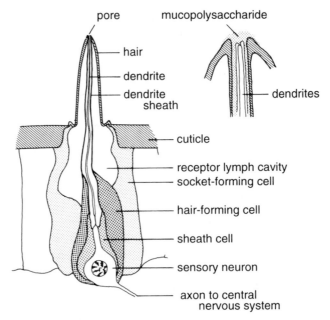

Figure 3.2. Diagram of the structure of a contact chemoreceptor. Only one neuron and dendrite is shown, but usually the number is four or five. Inset shows the tip of the hair in greater detail. The tips of the dendrites are bathed in a mucopolysaccharide and molecules must pass through this in order to stimulate the dendrites.

A single cell may have more than one kind of acceptor site, and this may usually be the case. In the sugar sensitive cells of flies there are three different kinds of acceptor molecules, presumably proteins, that interact with different types of "sugar" molecule. One of these is called a pyranose site because it reacts with glucose and arabinose and other pyranose sugars which have a six-membered ring. Sucrose, which is a major phagostimulant for phytophagous insects (see Section 4.2.4.1), also reacts with the pyranose site. A second class of acceptor molecule reacts with furanose sugars with a five-membered ring, like fructose and galactose. These are called furanose sites. The third class of site reacts with carboxylate anions of amino acids (Fig. 3.3). The possibility exists that there is even a fourth type of acceptor site that reacts with molecules having other characteristics.

The possession of different accceptor proteins by a single cell has important implications for the information an insect receives. Sugars that interact with the pyranose site, like glucose and arabinose, will compete for the same acceptor sites so that mixtures of the two sugars may be less than additive in their effects, depending on the concentrations. On the other hand, if glucose and fructose are

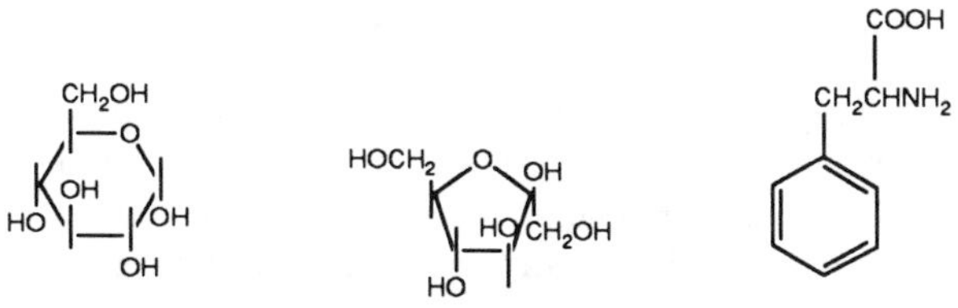

Pyranose sugar glucose | Furanose sugar fructose | Amino acid phenylalanine

Figure 3.3. Molecules that react with different acceptor sites on the dendrite of a sugar-sensitive cell of a fly.

present together their effects will be additive, irrespective of the concentrations, because they react with different acceptor sites. It is important to appreciate that the presence of different acceptor sites in a single cell does not provide the insect with information that enables it to discriminate between the stimulating molecules. Any stimulus generates a receptor potential and this gives rise to action potentials. The individual action potentials are the same whatever the stimulus, although the number may vary.

Cells responding to plant secondary compounds that elicit a deterrent behavioral response (see section 4.2.4.3) are often called deterrent cells. The deterrent cell in a single sensillum may respond to compounds in many different chemical classes, including even synthetic compounds in novel chemical classes. This presents a problem in terms of how these compounds affect the sense cells. Does the cell possess many different types of acceptor sites? An alternative possibility is that the deterrent compounds act by interacting nonspecifically with the cell membrane, not with specific acceptor sites. However, if this is the case, we should expect all sense cells, whatever their specific tuning, to be affected in a similar way. This is not the case. The method by which deterrent compounds initiate receptor potentials remains a mystery.

Cells responding to host-specific compounds are known to occur in a few species. The larva of *Pieris brassicae* has cells responding to glucosinolates in the sensilla of the galea, and similar cells are present in tarsal sensilla of the adult. Glucosinolates are characteristic of the cabbage family (Brassicaceae) on which these adults oviposit and the larvae feed. A cell responding to hypericin is present on the tarsal receptors of the beetle, *Chrysolina brunsvicensis*. Hypericin is a compound that is characteristic of St. John's wort (family Hypericaceae), the host plant of this insect. In both these cases these cells are believed to respond only to the host-specific stimuli, and not to other compounds. Other examples of responses to host-specific chemicals in monophagous and oligophagous insects are known, but such specificity has not been found in all insects with a restricted host-plant range. For example, no response to compounds specific to the Solanaceae has been found in the Colorado potato beetle, *Leptinotarsa decemlineata*, despite the fact that research has been directed towards finding such receptors.

3.2.1 Distribution and abundance of contact chemoreceptors

Contact chemoreceptors are usually present on the mouthparts, but they also occur on the antennae and tarsi (feet) of most insects and on the ovipositors of females. In addition, in Orthoptera, contact chemoreceptors are present on all parts of the body. In many insects, a few contact chemoreceptors are also found on the wings.

The numbers of chemoreceptors associated with the mouthparts vary greatly among taxa. Caterpillars commonly have two sensilla on the inner, epipharyngeal, face of the labrum, two on each galea, eight on each maxillary palp, and three on each of the antennae (Fig. 3.4). Some of those on the palp and all three on the antennae are olfactory receptors; the rest are contact chemoreceptors. Table 3.1 shows the chemicals to which some of the cells associated with these sensilla respond in one species. All species of caterpillar, irrespective of their feeding habits, have similar numbers of chemosensitive sensilla on the mouthparts, and different instars have similar numbers. Contact chemoreceptors are not known to occur on other parts of the body of caterpillars. By contrast, in adult Lepidoptera, contact

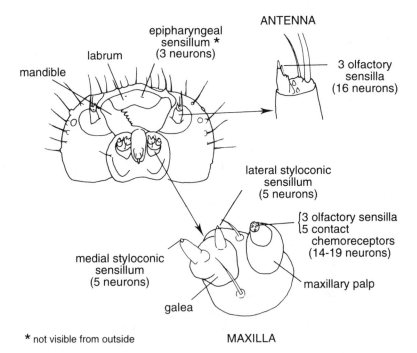

Figure 3.4. Diagram of the head of a caterpillar seen from below with enlargements of an antenna and a maxilla. Chemosensory sensilla are labelled and the number of neurons associated with each group of sensilla is given (partly after Schoonhoven, 1987).

Table 3.1. Main classes of compounds to which some of the sensory neurons on the mouthparts of Pieris brassicae respond (partly after Schoonhoven and Blom, 1988).

Sensillum	Cell 1	Cell 2	Cell 3	Cell 4	Cell 5
Lateral styloconic on galea	amino acids	glucose, sucrose	mustard oil glucosides	deterrents	mechanical stimuli
Medial styloconic on galea	inorganic salts	many sugars	mustard oil glucosides	many deterrent compounds	mechanical stimuli
Epipharyngeal	inorganic salts	sucrose	—	deterrents	—
Mandibular canals	—	—	—	—	mechanical stimuli

chemoreceptors are present on the tarsi and, in females, on the ovipositor, in addition to those on the mouthparts. There is usually a relatively large number, >100, on the proboscis, as well as receptors in the wall of the cibarial (preoral) cavity and on the labial palps. The pattern of response of the receptors of adult moths is generally similar to that in caterpillars of the same species (Fig. 3.5a) and adult tarsal and proboscis sensilla also respond in similar ways, although the proboscis sensilla tend to be more sensitive to phagostimulants, producing more action potentials in response to a given stimulus (Fig. 3.5b).

Plant-sucking bugs also have small numbers of contact chemoreceptors, primarily on the tip of the labial sheath that surrounds the stylets and also in the wall of the cibarial cavity. Surprisingly, aphids have only mechanoreceptors (touch receptors) at the tip of the proboscis and their only contact chemoreceptors on the mouthparts are within the cibarial cavity.

In grasshoppers and other Orthoptera, the picture is quite different. The mouthparts have very large numbers of contact chemoreceptors which are arranged in groups as well as being scattered over the whole surface (Fig. 3.6). This figure also shows the numbers of contact chemoreceptors and associated sensory neurons on the mouthparts of an adult of *Locusta migratoria*. It has over 16,000 sense cells in the main sensillum groups on the mouthparts. Unlike the caterpillars, the number of sensilla increases with the size of the insect; in the first instar nymph of *Locusta migratoria* there are only about 3,000 neurons. In addition, there are differences in the numbers of sensilla in species with different feeding

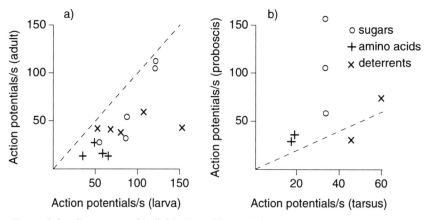

Figure 3.5. Responses of individual sensilla of *Helicoverpa armigera* to stimulation with a range of sugars, amino acids and deterrent compounds. **a)** Comparison of the firing rate of sensilla on the adult proboscis with that of sensilla on the larval galea. Dashed line represents equal firing rates by the two sensilla. **b)** Comparison of the firing rates of sensilla on the proboscis and tarsi of the adult (after Blaney and Simmonds, 1988, 1990).

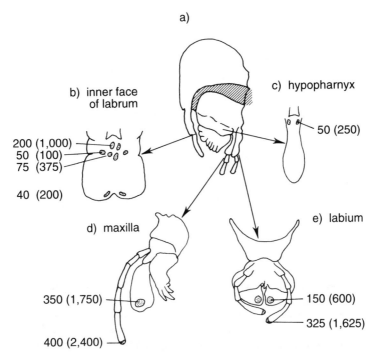

a)

b) inner face
of labrum

c) hypopharnyx

50 (250)

200 (1,000)
50 (100)
75 (375)

40 (200)

d) maxilla

e) labium

350 (1,750)

150 (600)

325 (1,625)

400 (2,400)

Figure 3.6.   Diagram of the head of a grasshopper showing the positions of the groups of chemoreceptors. **a)** A saggittal section of the head showing the positions of the mouthparts. **b-e)** Individual mouthparts showing the surfaces that come into contact with the food as it passes towards the mouth. The position of each of the major sensillum groups is indicated by a shaded area. This distribution is found in all grasshoppers, but the numbers in each group vary. In this diagram, the number adjacent to each group shows the approximate number of sensilla in that group and, in brackets, the approximate number of chemosensory neurons in the group in an adult of *Locusta migratoria*. The total exceeds 16,000 neurons.

habits. Insects that feed on grasses usually have fewer sensilla in some of the sensillum groups than insects of similar size that feed on a wide range of plants. Even within a subfamily, species that are more restricted in their food range tend to have fewer sensilla in some of the groups (Fig. 3.7). Relatively little work has been done on the physiology of grasshopper contact chemoreceptors, but what there is indicates that the cells are less specific than those in Lepidoptera and the cells that respond to sucrose also respond to sodium chloride. Grasshoppers do have a deterrent cell responding to some plant secondary compounds. Sensilla on the legs appear to respond in the same way as sensilla on the maxillary palps, although specific comparisons have not been made.

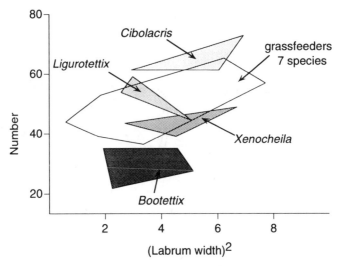

Figure 3.7. Numbers of sensilla in the most distal group of the labrum of different grasshoppers (see Fig. 3.6) in relation to the size of the insect. The square of the labrum width is used as a measure of size. Each polygon shows the results from one or a group of species. *Cibolacris* is polyphagous, *Ligurotettix* is oligophagous, and *Bootettix* and *Xenocheila* are monophagous. All these grasshoppers belong to one subfamily, Gomphocerinae, most of which are oligophagous on grasses (after Chapman and Fraser, 1989).

3.2.2 Variability of response

Although it used to be thought that the response from a given receptor to a given stimulus was constant, it is now known that the sensitivity of a single contact chemoreceptor to a single compound may be highly variable. Most of this work has been on the caterpillar of *Spodoptera littoralis* and the final instar larva of *Locusta migratoria*.

The sensitivity varies through the instar and over the course of a day; in *S. littoralis*, it is generally higher in the morning than in the afternoon (Fig. 3.8). Nerve cells responding to different chemicals do not necessarily change in the same way even in the same sensillum. In addition to these temporal changes, changes also occur in relation to feeding and nutritional status. In *Locusta migratoria* the tips of the sensilla on the maxillary palps close after a meal so that these sensilla generally fail to respond to stimulation. The pores reopen after one to two hours. After feeding on an artificial diet containing a high level of protein, the sensory response to amino acids is reduced; similarly, the response to sucrose is suppressed after feeding on artificial diet containing high sugar levels (Fig. 3.9).

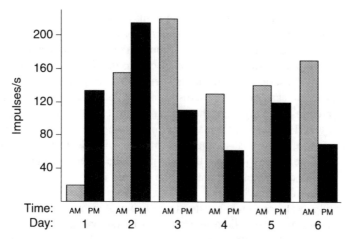

Figure 3.8. Temporal variation in the response of the lateral sensillum on the galea of the larva of *Spodoptera littoralis* (see Fig. 3.4) to stimulation with 0.05M sucrose. AM=morning. PM=afternoon, 1-6=days in the final larval instar (after Simmonds et al., 1991).

These variations in sensory response correlate well with behavioral changes. Food intake reaches a peak in mid-instar; *S.littoralis* eats more in the morning than in the afternoon; and feeding is suppressed after a meal. Showing a preference for a nutrient previously in short supply (see section 5.6.1) is also associated with changes in the receptors. There is as yet, however, no good evidence to show whether the sensory changes cause the differences in food intake or whether they merely accompany them.

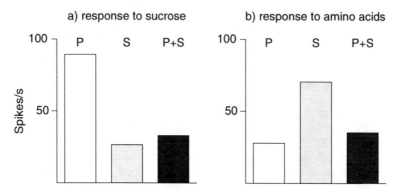

Figure 3.9. Responses of sensilla on the maxillary palp of *Locusta migratoria* after feeding for four hours on articifial diets containing protein, but no carbohydrate (P), digestible carbohydrates but no protein (S), and with both present (P+S). The sensilla were stimulated with sucrose or a mixture of amino acids equivalent to 0.01M (after Simpson et al., 1991).

Another type of variation in sensory input may occur in the course of feeding as a result of interactions within the sensillum. For example, in the tarsal sensilla of adult *Helicoverpa armigera*, sucrose stimulates one cell, while azadirachtin— a deterrent compound—stimulates another cell in the same sensillum. Sucrose alone does not cause activity in the cell responding to azadirachtin, nor does azadirachtin cause activity in the sucrose-sensitive cell. However, when the two chemicals are presented together, they each have the effect of depressing activity in the other cell (Fig. 3.10a). Similar experiments with the grasshopper, *Schistocerca americana*, show that as the concentration of a deterrent compound is increased, not only does it cause the activity of the deterrent-sensitive cell to increase, it also completely suppresses the activity of the cells responding to sucrose (Fig. 3.10b).

These findings are important because they show that the sensory input obtained by an insect when it feeds on a leaf may not be equivalent to the sum of the inputs from single chemicals. In other words, the nature of the chemical mixture

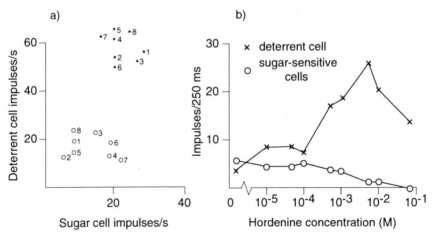

Figure 3.10. Interactions between neurons within a sensillum. **a)** Responses of tarsal sensilla of *Helicoverpa armigera* when stimulated with sucrose or azadirachtin. Each point represents the response of one sensillum (1-8). Sucrose stimulated the sugar cell, but not the deterrent cell; azadirachtin stimulated the deterrent cell, but not the sucrose cell. Dots show the responses of these cells when each sensillum was stimulated with each of the two chemicals independently. When the chemicals are applied at the same concentrations in a mixture, the firing rates of both cells are reduced (open circles) (after Blaney and Simmonds, 1990). **b)** Responses of sensilla on the tibia of a grasshopper, *Schistocerca americana*, to stimulation with a mixture of sucrose and hordenine. As the hordenine concentration was increased, the firing rate of the deterrent cell increased and remained high, but the response to sucrose decreased and finally disappeared altogether. The sucrose concentration was constant throughout the experiment (after Chapman et al., 1991).

in the leaf is important. They also show that behavioral deterrence may result not only through the activity of the deterrent cells, but also through the suppression of activity of cells responding to phagostimulants. Sometimes, the deterrent effect seems to be produced only by suppression of these cells since the deterrent cell does not fire. This is the case with the nonprotein amino acid canavanine in the grasshopper.

3.2.3 Sensory coding

The behavioral response of an insect is a consequence of the various sensory inputs from the food. Although we have no direct experimental evidence concerning the ways in which the inputs from contact chemoreceptors are integrated, there is some evidence from caterpillars which provides the basis for a hypothetical scheme. In *Pieris brassicae* there is a direct relationship between the amount eaten and the activity of the receptor cells in response to different concentrations of sucrose and sinigrin (Fig. 3.11a). Conversely, as the activity of the deterrent cell increases with the concentration of deterrent, the amount eaten declines (Fig. 3.11b). It is presumed that these inputs are brought together in the central nervous system in an additive manner, sucrose and sinigrin having positive effects while deterrents have negative effects. If the positive inputs outweigh the negative

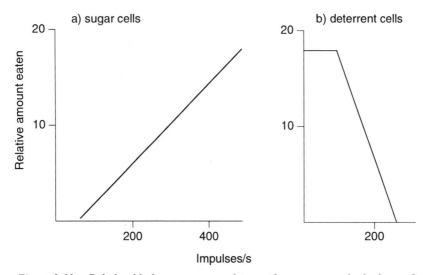

Figure 3.11. Relationship between sensory input and amount eaten in the larva of *Pieris brassicae*. **a)** Amount eaten in relation to the combined activities of the three sugar-sensitive cells when stimulated by sucrose (Table 3.1). **b)** Amount eaten in relation to the combined activities of two deterrent cells when stimulated by strychnine in the presence of 0.02M sucrose. Amount eaten was based on the production of feces (after Schoonhoven and Blom, 1988).

inputs, the insect feeds; if the converse is true, it does not (Fig. 3.12). It seems very likely that some similar system operates in all the insects.

The balance between positive and negative influences suggested in Fig. 3.12 will vary, not only as a result of differences in the quality of the stimulus and the state of repletion of the insect, but also as a result of variation in the responsiveness of the receptors as outlined above. In addition, there may be short-term changes in the balance of inputs because of the different physiological characteristics of the receptor cells. It seems to be generally true that cells that are stimulated by phagostimulatory chemicals adapt more quickly than deterrent cells (Fig. 3.13). Consequently, the balance of positive and negative inputs will change even in the course of a single period of stimulation, with the negative effect assuming preponderance over time.

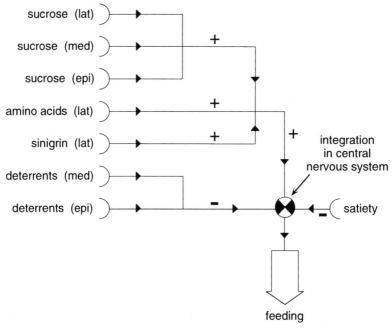

Figure 3.12. Schematic representation of how the inputs from different mouth-part receptors might be integrated within the central nervous system to regulate feeding in the caterpillar of *Pieris brassicae*. Inputs from the sucrose, amino acid and sinigrin cells in the lateral (lat) and medial (med) sensilla on the galea and those on the epipharynx (epi) would have positive effects (+) tending to induce feeding. Inputs from the deterrent cells would have negative effects (-) tending to inhibit feeding. The degree of satiety of the insect would also be important, tending to inhibit feeding when the gut was already full. "Feeding" or "not-feeding" results from the balance of these positive and negative inputs. Compare Figs. 3.4 and 3.11 and Table 3.1 (after Schoonhoven, 1987).

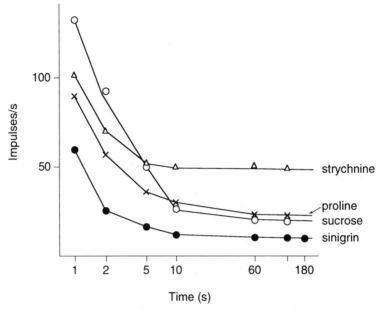

Figure 3.13. Adaptation rates of different sensory neurons in the caterpillar of *Pieris brassicae*. The deterrent cell, which is stimulated by strychnine, maintains a higher firing rate with continuous stimulation than the other three cells which have positive effects on feeding (after Schoonhoven and Blom, 1988).

This type of sensory processing apparently does not involve complex analysis in the central nervous system, but simply the integration of the various inputs. Each sensory axon carries unambiguous information: feed or do not feed. This is called a labelled line system. An alternative is known to occur in vertebrates and perhaps also occurs in insects. Here the individual sensory cells respond to chemicals in different chemical classes. Different cells vary in their relative sensitivities to these chemicals and their response spectra overlap. In this case, the information carried by one axon does not convey a clear message; it requires the central nervous system to interpret the message from the totality of the sensory input. This type of integration is called across-fiber patterning. Since most receptor cells respond to some range of compounds, even if the range is limited, it may be argued that decision-making by insects nearly always involves an element of across-fiber patterning. We shall not be able to answer this question until we have a much more complete understanding of the physiology of the insect central nervous system.

Unlike the olfactory system (see below), there is no known center in the central nervous system which serves as a focus for all the inputs from contact chemoreceptors. The axons from most of the sensilla on the mouthparts extend to the subesophageal ganglion, but the arborizations of axons from sensilla on the maxillae and labium remain separate in their distinct parts of the ganglion (neuromeres). Chemoreceptors on the legs have axons ending in the corresponding segmental ganglion. So we do not know where the various inputs depicted in Fig. 3.12 are finally integrated.

3.3 The sense of smell

The cuticle of olfactory sensilla is perforated by numerous small pores, ranging from about 10 to 50 nm in diameter in different insects. This is called the multiporous condition. In some cases the cuticular structure is hairlike, in others it is a flat plate perforated with pores. These are called pore plates or plate organs. Plate organs are common in Hymenoptera, some Coleoptera and Homoptera. In other cases a very short sensory hair, or peg, is sunk in a cavity within the cuticle; these are called coeloconic sensilla. In other respects, olfactory sensilla resemble contact chemosensitive sensilla except that the dendrite sheath ends at the base of the hair. The position of this ending is sometimes visible externally. The dendrites leave the sheath at the base of the hair and extend up into the cavity of the hair, passing close beneath the cuticular pores (Fig. 3.14).

The number of sensory cells in each sensillum varies. Sensilla containing nerve cells that respond to plant odors often have two or three sensory neurons,

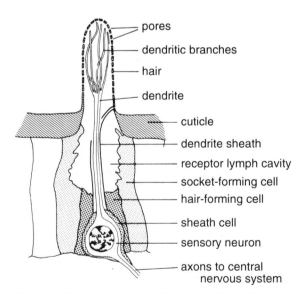

Figure 3.14. Diagram of an olfactory sensillum. The size of the pores is greatly exaggerated and their density greatly reduced compared with the true situation.

but there may be many more. In grasshoppers, for example, there are two morphologically distinguishable types of olfactory hair, one with only three nerve cells and one with as many as 30. Grasshoppers also have coeloconic sensilla with three sense cells. In some cases the dendrites branch at the base of the hair after exiting the dendrite sheath.

Olfactory cells often produce action potentials spontaneously even in the absence of any deliberate stimulation, and stimulation of the cell by an appropriate compound produces a change in this background rate of firing. In general, stimulating odors produce an increase in the firing rate, but examples are known where stimulation causes a reduction in the level of activity.

Phytophagous insects can smell a wide range of odors. Their ability to smell is often demonstrated by recording from the whole antenna to produce an antennogram. This is done by inserting an electrode into each end of the antenna, either while it is still attached to the insect, or immediately after its removal. The potential difference between the two electrodes is recorded and when a puff of air carrying an odor is blown across the antenna the potential changes. The change in potential is probably a measure of the summed receptor potentials of all the nerve cells in the antenna that respond to the odor and it is known that the amplitude of the electroantennogram (EAG) is proportional to the number of sensilla present. More precise data are obtained by recording from single olfactory cells.

All the leaf-feeding insects that have been examined critically have been shown to be able to smell components of the commonly occurring green leaf volatiles such as hexanol and hexenal (see Section 2.1). This applies to insects that feed directly on leaves, like grasshoppers, aphids and beetles, and to those that lay their eggs on leaves, like adult Lepidoptera. Flies whose larvae feed on the roots or shoots also respond to these compounds. It has also been shown in a variety of insects that the number and/or sensitivity of receptors is greater for alcohols and aldehydes with six-carbon-atom chain lengths, that are major constituents of the "green odor," than for compounds with shorter or longer chains. Not only the chain length, but the precise form of the molecule determines sensitivity (Fig. 3.15).

Because of this general sensitivity, all phytophagous insects probably have the capacity to smell any plant, whether it is a host or not. Fig. 3.16 shows the electroantennogram responses of four related grasshoppers to the odors of a range of host and nonhost plants. *Chorthippus curtipennis* is graminivorous, feeding only on *Aristida* and *Cynodon* of the plants tested. *Ligurotettix coquilletti* is oligophagous, and would feed on *Larrea*, *Atriplex* and *Fagonia* of the plants tested. *Bootettix argentatus* is monophagous on *Larrea*, and *Cibolacris parviceps* is polyphagous. It would eat all of the plants except *Simmondsia*. Despite these differences, all four species could smell all the plants and exhibited similar spectra of responses to them. Responses to the grasses were no different in the grass-feeder than in the other species and the biggest EAG in all the species was

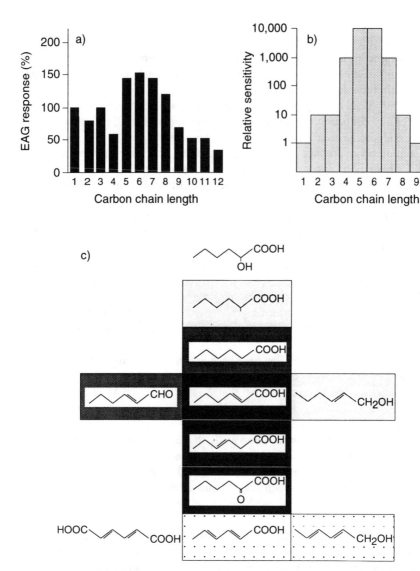

Figure 3.15. Sensitivity of receptors to "green-odor" components. **a)** Sensitivity of receptors on the antennae of female fruit flies, *Dacus dorsalis*, to primary alcohols with different chain lengths. Electroantennogram (EAG) responses expressed as a % of standard (after Light and Jang, 1987). **b)** Sensitivity of single cells in coeloconic sensilla of *Locusta migratoria* to volatile organic acids (after Kafka, 1971). **c)** Sensitivity of single cells in coeloconic sensilla of *Locusta migratoria* to a range of compounds with chain lengths of 6 carbon atoms. Intensity of shading shows the strength of the response; unshaded indicates no effect (after Kafka, 1971).

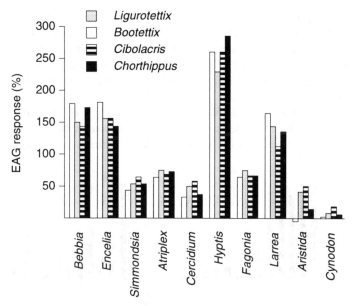

Figure 3.16. Electroantennogram (EAG) responses of four Gomphocerine grasshoppers to a range of host and nonhost odors. Notice the
general similarity of their response regardless of the feeding pattern of
the insect. EAG responses are expressed as a % of standard. *Bootettix
argentatus* is monophagous on *Larrea*. *Ligurotettix coquilletti* is oligophagous; it feeds on *Atriplex*, *Fagonia* and *Larrea*. *Cibolacris parviceps* is polyphagous; it eats all the plants except *Simmondsia*. *Chorthippus curtipennis* is oligophagous on grasses (*Aristida* and *Cynodon*)
(after White and Chapman, 1990).

produced by *Hyptis*, desert lavender, which smells most strongly to the human
nose, but is not eaten by three of the species. Such results do not necessarily
mean that the insects are unable to distinguish hosts from nonhost by olfaction,
since the EAG does not give any idea of the specificity of the cells responding.

In the same way that leaf-eating insects respond to general leaf odors, insects
that feed in the wood or bark of pine trees respond to a range of chemicals
produced by these trees in their resins. The best known of these are α-pinene
and camphene.

Despite these general responses observed when we look at the activity of the
antenna as a whole, the response spectra of individual olfactory cells in any
insect may vary considerably. For example, in *Leptinotarsa decemlineata* the
cells responding to the green leaf volatiles can be grouped into 4 classes according
to the strengths of their responses to different components. Classes A and B
respond to a number of different compounds, but classes C and D are more

specific (Fig. 3.17). A fifth class of cell (E) is insensitive to all the green leaf volatiles, but responds strongly to methylsalicylate. This apparent specificity may be an artefact of the range of compounds used in the experiments, but clearly indicates the presence of some cells in the olfactory sensilla with quite different response spectra from the majority. Class E cells constituted only 4% of the total tested.

In addition to these responses to widely occurring plant volatiles, some insects also exhibit sensory responses to the odors of compounds that are specific to their host plants. The onion fly, *Delia antiqua*, exhibits a large EAG response

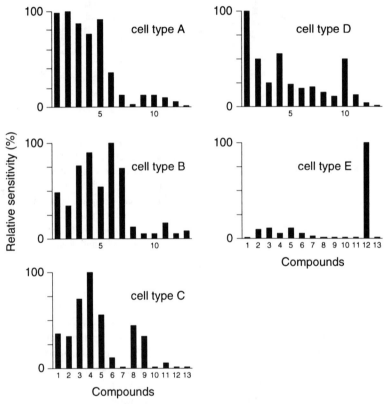

Figure 3.17. Relative sensitivities of neurons in olfactory sensilla of *Leptinotarsa decemlineata* to a range of different compounds. The cells are grouped into five classes (A-E) according to their pattern of response. Sensitivity is expressed relative to the maximum response for each group of cells. Compounds are: 1, *cis*-3-hexen-1-ol; 2, *cis*-2-hexen-1-ol; 3, *trans*-3-hexen-1-ol; 4, *trans*-2-hexen-1-ol; 5, n-hexanol; 6, hexanol-2; 7, hexanol-3; 8, *trans*-2-hexenal; 9, hexanal; 10, *cis*-3-hexenyl acetate; 11, linalool; 12, methylsalicylate; 13, α-pinene (after Ma and Visser, 1978).

to some of the volatile compounds from the onion, and *Psila rosae*, the carrot fly, has a large EAG response to carrot volatiles (Fig. 3.18). These results show that there are many cells responding to these compounds in the antennae. Propyldisulfide (compound 5 in Fig. 3.18) is an attractant for *D. antiqua* in the field, and the same is true of *trans*-asarone (compound 11) for *P. rosae*. The picture is different, however, for *Delia brassicae*, the cabbage rootfly. This insect does not exhibit a large EAG response to allylisothiocyanate, one of the characteristic cabbage volatiles, and some of the compounds produced by leeks, which are not hosts for this species, cause a bigger EAG. Nevertheless, *D. brassicae* is attracted by the odors of cabbage plants and by allylisothiocyanate alone. This apparent anomaly highlights the fact that the results of EAG studies must be interpreted with caution. A large EAG indicates that many neurons are responding to the odor; a small one shows that only a few cells are responding. But if these cells are highly specific, the information they convey to the nervous system may nevertheless be sufficient to elicit a response. The size of the EAG is not the only factor to consider.

Behavioral evidence suggests that cells responding to specific host-related compounds also occur in other insects. For example, the aphid *Cavariella aegopodii*, is attracted by water traps baited with carvone and it is likely that the insect has receptors that are specific for this compound. An isomer of carvone occurs in some of the insect's hosts. Allylisothiocyanate, which is characteristic of the Brassicaceae, attracts a number of *Brassica*-feeding species in addition to *Delia brassicae*, with the implication that these species have sense cells that are specific to this substance.

However, it is not necessarily always true that insects have receptors for chemicals specific to their hosts. For example, cells responding to specific host-plant odors have not been found, either in adult *Pieris brassicae,* or in *Leptinotarsa decemlineata*. These species can, nevertheless, recognize their hosts olfactorily as explained in section 4.1.1.

As with contact chemoreception, it is assumed that the sensitivity of the receptors depends on the presence of acceptor sites in the dendrite membranes of the sense cells. Cells responding to a range of compounds presumably have a number of different types of acceptor sites, while more specific cells have only a small number of types.

3.3.1 Distribution and abundance of olfactory receptors

In most insects all, or nearly all, the olfactory receptors occur on the antennae. However, small numbers are present on the palps in grasshoppers, and on the labial palps of some Lepidoptera.

The number of olfactory sensilla on an antenna of an insect may be very large, up to 100,000 pheromone receptors are present on the antennae of some male moths and other insects. This large number probably reflects the low concentra-

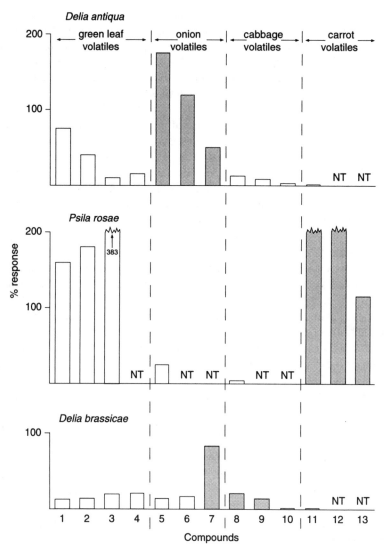

Figure 3.18. Electroantennogram responses of three flies to volatiles from their hosts and nonhosts. Specific host-plant chemicals are hatched. Note that dimethyldisulfide (compound 7) is produced by cabbage as well as by onion. Responses are shown relative to the response to *cis*-3-hexen-1-ol at 10 μlml^{-1}. Compounds are: 1, hexanol; 2, *trans*-2-hexen-1-ol; 3, hexanal; 4, hexylacetate; 5, propyldisulfide; 6, dipropyldisulfide; 7, dimethyldisulfide; 8, allylisothiocyanate; 9, tetrabutylisothiocyanate; 10, benzylisothiocyanate; 11, *trans* asarone; 12, *trans*-methyl-iso-eugenol; 13, eugenol. NT = not tested (after Guerin and Städler, 1982; Guerin and Visser, 1980; Guerin et al., 1983).

tions in which pheromones occur and to which the insect must respond. Plant volatiles are generally present in higher concentrations, and so the insects probably need fewer sensilla to perceive a host odor at some appropriate distance. Nevertheless, the numbers of olfactory sensilla concerned with the detection of plant odors may be very large. In adult *Locusta migratoria*, for example, there are about 3,000 hairlike olfactory receptors and 2,000 coeloconic sensilla on each antenna. In the male of *Manduca sexta* there are probably about 550 olfactory sensilla on each annulus of the antenna that are not concerned with pheromone perception; this amounts to about 44,000 on the antenna as a whole. Much smaller numbers are present on small insects like aphids. For example, winged females of *Acyrthosiphon pisum* have 24 plate organs and four coeloconic pegs on each antenna. Apterous females have only nine plate organs, and larval aphids even fewer.

By comparison with the adults, larvae of holometabolous insects generally have only small numbers of olfactory sensilla. Thus the caterpillar of *Manduca sexta* has three olfactory sensilla on each antenna and five on each maxillary palp. This compares with the thousands present in adults. Although direct comparisons within species are not possible amongst the phytophagous beetles, beetle larvae that have been examined have fewer than 10 putative olfactory receptors on the antennae.

The length of the antenna is proportional overall to the number of olfactory receptors which it bears, but this relationship does not hold for small fragile hemipteroids which have fewer sensilla than would be expected for their size (Fig. 3.19). The general relationship possibly reflects the fact that small insects are relatively weak fliers and cannot make headway against winds exceeding about 1.5 m s^{-1}. A large number of sensilla implies a high level of sensitivity and this might cause the insect to become activated in wind conditions where, if it is very small, it would be unable to reach the source of the odor.

3.3.2 Integration of olfactory inputs

The axons from olfactory sense cells all end in an area of the brain called the antennal lobe. This lobe has a series of discrete regions of neuropile called glomeruli. These are the primary centers where extensive integration of the olfactory sensory input occurs. For example, in *Locusta migratoria* more than 50,000 sensory axons converge onto about 1,000 glomeruli and about 800 output axons leave the olfactory lobe. In adult *Manduca sexta* the degree of convergence is even greater. Over 150,000 axons from cells that are sensitive to plant odors converge on to about 60 glomeruli. It is probable, though unproven, that different glomeruli are the centers in which the inputs from sense cells with similar response characteristics come together. One advantage of such extensive convergence is that the incoming signal is greatly amplified, and the signal-to-noise ratio is improved.

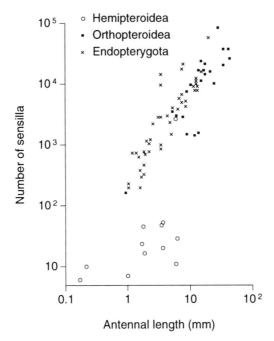

Figure 3.19. Relationship between antennal length and number of olfactory sensilla (after Chapman, 1982).

3.4 The sense of touch

A number of different systems are involved in the mechanosenses of insects. In this section we will consider only those that appear to be involved more or less directly in plant/insect interactions.

The sense of touch is mediated by hairs. When the response of the hair is purely tactile, the hair has no pores, and it is classified as aporous. A single dendrite ends at the base of the hair and it is characterized by the possession of a dense bundle of microtubules, called the tubular body, at the tip (Fig. 3.20). Movement of the hair in its socket distorts the tubular body and initiates the production of the receptor potential. In hairs concerned with touch, it is likely that stimulation occurs only during movement of the hair; sustained deflection does not produce a sustained response. However, in hairs concerned with proprioception, the response may continue as long as the hair is deflected. In many cases, contact chemoreceptor hairs also function as mechanoreceptors. These have a mechanosensitive dendrite with a tubular body ending at the base of the hair, in addition to the chemosensitive dendrites that extend to the tip. In some cases, mechanosensitive hairs are known to be directionally sensitive, responding

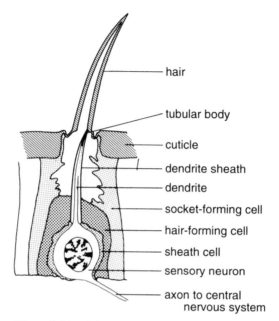

Figure 3.20. Diagram of a mechanoreceptor.

more strongly when moved in one direction. It is not known if this is true for most of the hairs concerned with touch.

In addition to touch, insects respond to the hardness of food. This is probably measured by sensilla in the heavily sclerotized cusps of the mandibles, and sometimes also the maxillae. These sensilla consist of pairs of sense cells with dendrites extending into a canal which passes almost through the cuticle, ending just beneath the surface (Fig. 3.21). The dendrites do not possess tubular bodies, but at the lower levels beneath the cuticle they are surrounded by a fibrous structure called a scolopale. These receptors belong to a class called chordotonal organs. They are stimulated by pressure on the mandibular cusps, but the mechanism by which this produces a receptor potential is not known.

3.4.1. Distribution of touch receptors

Hairs which respond to touch are widely distributed over all parts of the body. Axons from the receptors arborize in the relevant segmental ganglion of the central nervous system; there is no central area in the nervous system where all information on touch is integrated. It is obviously essential that touch receptors retain their individual identity when their axons reach the central nervous system, since, if this were not the case, the insect would be unable to determine which part of its body was making the contact. Reflecting this, we know that there is at least some degree of spatial mapping within each ganglion. In the grasshopper,

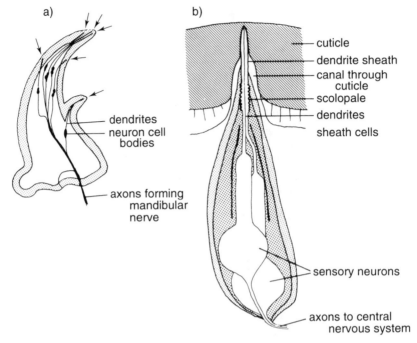

Figure 3.21. Diagrams of the canal sensilla in the mandibles of a beetle larva. **a)** Positions of canal sensilla (arrows) on the mandible. **b)** Structure of a single sensillum. All the cells are much more elongate than shown in the diagram. A third, inner sheath cell is omitted (after Zacharuk and Albert, 1978).

the arborizations of axons from hairs on the tarsus are more lateral in the ganglion than those of hairs on the tibia which in turn are more lateral than the arborizations of femoral hairs.

3.5 The sense of sight

The principal organs of vision in adult insects and larval hemimetabolous insects (grasshoppers and sucking bugs) are the compound eyes. They consist of large numbers of similar units, called ommatidia. In the locust, *Schistocerca gregaria*, there are about 2,500 ommatidia in each eye of a first instar larva, and about 9,500 in the adult.

Each ommatidium is made up of a light-gathering part and a sensory part. All over the surface of the eye the cuticle is transparent. It is developed into a series of, usually hexagonal, closely packed lenses, each corresponding with an ommatidium. Beneath each cuticular lens is a second lens, the crystalline cone (Fig. 3.22).

Beneath the crystalline cone is the receptor system. This typically is formed from eight elongate retinula cells which, along the longitudinal axis of the ommatidium, are produced in densely packed microvilli. The array of microvilli bordering each retinula cell is called a rhabdomere. In most insects other than Diptera, the rhabdomeres of all the retinula cells are closely juxtaposed. They are collectively called the rhabdom. At the base of the eye short axons from each retinula cell pass back to a region of integration called the lamina ganglionaris in the optic lobe of the brain.

Each ommatidium is surrounded by a series of pigment cells which can prevent light that passes obliquely through the eye from entering the ommatidium. This screening pigment is quite different chemically and functionally from the visual pigment present in the retinula cells.

The visual pigment, retinal, is present in the membrane of the microvilli of the retinula cells. This compound, derived from vitamin A, which must therefore be present in the diet, is linked with a protein called a chromoprotein. Light falling on the eye causes the pigment to isomerize and this initiates a chain of chemical events which produce a change in membrane permeability, and so lead to the development of the receptor potential. The eye differs from other sensory systems in that this receptor potential does not give rise to action potentials in the axon of the sensory cell. Instead, the graded receptor potential spreads along

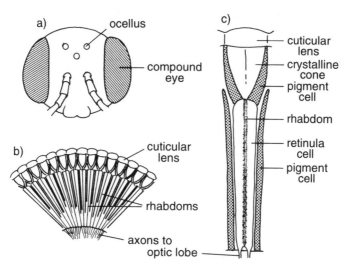

Figure 3.22. Compound eye. **a)** Head of an insect showing the compound eyes and ocelli. **b)** Diagram of a section though a compound eye. **c)** Diagram of a single ommatidium.

the axon resulting in the release of synaptic transmitter substance where the axon terminates in the lamina ganglionaris or even more proximally in the optic lobe.

The range of wavelengths of light absorbed by the visual pigment depends on the nature of the chromoprotein with which the retinal is associated. Insects commonly have three different visual pigments, one with a maximal sensitivity in the ultraviolet, one with a peak of sensitivity in the blue part of the spectrum, about 450 nm wavelength, and a third with a peak at about 540 nm, in the green region (Fig. 3.23). Each retinula cell possesses only a single pigment, and the separate retinula cells of an ommatidium have different pigments.

The possession of pigments in separate cells absorbing different ranges of wavelengths is essential for color vision. The ability to see ultraviolet differentiates most insects from humans, although our inability to see ultraviolet results from filtering in the lens system rather than from the lack of an appropriate visual pigment. On the other hand, the lack of a pigment with sensitivity to longer wavelengths means that insects cannot, in general, distinguish reds. Consequently the patterns seen by many insects will be somewhat different from those visualized by humans.

However, many butterflies and moths are able to distinguish red as a color. This is due to the presence of a visual pigment with maximum absorption of light with wavelengths around 600 nm. This pigment is additional to the three responding to the shorter wavelengths and these insects have four different visual pigments. In a single ommatidium, the moth *Spodoptera exempta*, has six green-sensitive cells, one red-sensitive cell and one cell with maximum sensitivity either in the ultraviolet or in the blue regions.

There has been relatively little critical study of the ability of phytophagous

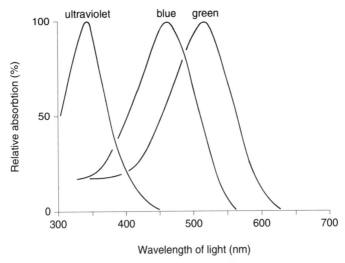

Figure 3.23. Color vision. Absorption of light of different wavelengths by the three visual pigments commonly present in insects.

insects to discriminate colors. It is certain that adult Lepidoptera can do so. In addition, studies on phytophagous species have demonstrated color vision in a grasshopper, some plant-sucking bugs and a beetle. Color vision has also been demonstrated in a number of fly species, although not categorically in any phytophagous species. The fact that an insect responds differently to different colors is not, by itself, good evidence that the insect has color vision since it is generally true that different colors have different brightnesses as well as reflecting different wavelengths (see section 4.1.2.2).

The compound eye is well adapted to perceive form. It is believed that in daylight, only light coming from a limited area in front of each ommatidium passes through the lens system and reaches the rhabdom. Light travelling more obliquely is absorbed by the screening pigment. Hence an ommatidium directed towards a dark area will receive less light than one directed towards a light area. Consequently any object, such as a plant, will appear at the back of the lens system as a series of points of light of differing intensity. The image has been likened to the appearance of a rather coarse-grained newspaper photograph. The points of light will stimulate the retinula cells according to their intensity and so it is presumed that the insect perceives the object.

The ability of the compound eye to resolve fine details depends on the angular separation between the receptors, the rhabdoms. Obviously if the eye has only a few ommatidia, each of which receives light from a relatively large area, resolution of an object can only be poor. On the other hand, if there are many ommatidia, each with a small visual field, the image will be much sharper (Fig. 3.24). In most insects the angular separation between the rhabdoms is 1–2°. Sometimes it varies in different parts of the eye, and this is indicated by the different sizes of the cuticular lenses. These higher levels of resolution are usually associated with predaceous insects, or insects that mate on the wing, rather than with phytophagous insects.

The inputs from the compound eye are integrated in the optic lobe which is situated immediately behind the eye, continuous with the brain. The optic lobe consists of three masses of axons known as the lamina ganglionaris and the medulla externa and interna. These occur at successively deeper levels in the optic lobe and are connected to each other by regions in which the axons cross over each other in a regular pattern. Many synapses occur in the axon masses. Axons connect the medulla interna with integrating centers in the brain.

Adult insects and larval hemimetabolous insects have, in addition to the compound eyes, three simple eyes, or ocelli. Since these are probably not involved in the insects' interactions with plants, they are not considered here.

Larval holometabolous insects (butterflies and moths, beetles and sawflies), do not have compound eyes, but often they have simple eyes which are known in some cases to have a role in responses to host plants. Caterpillars typically have six of these stemmata on each side of the head; larval sawflies have only one. They are also present in beetle larvae. In all cases, they comprise a lens

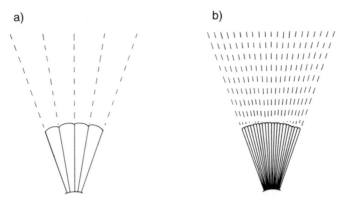

Figure 3.24. Visual acuity. The ability of the compound eye to resolve details depends on the angle from which light reaches the rhabdom. **a)** Each ommatidium receives light over an angle of 10°. Consequently, a narrow object whose long axis subtended an angle of 40° at the eye would be resolved by only four ommatidia. The ability to resolve detail is poor. **b)** Each ommatidium receives light from a 2° angle. A narrow object whose long axis subtended an angle of 40° at the eye would be resolved by 20 ommatidia. Resolution is much better. If the object is square with sides equal to the long axis of the narrow object it would stimulate 16 ommatidia in (a) and 400 ommatidia in (b).

system and a receptor system similar in basic structure to that in compound eyes. However, the details vary from species to species and these details determine the visual capabilities of the larva. In caterpillars, each of the stemmata has a distal and a proximal rhabdom aligned along the axis of the eye (Fig. 3.25a,b). Each of the stemmata can be compared with a single ommatidium and the ability of a caterpillar to resolve the shape of objects is limited by the fact that it has only six units on each side. Caterpillars thus do not have the capacity to perceive the structure of a plant.

By contrast, the single lens of a sawfly larva is backed by a large number of separate rhabdoms, each formed by eight retinula cells (3.25c,d). The angle between the rhabdoms varies between 4° and 6°, compared with 1–2° in adult insects. So sawfly larvae have the ability to resolve objects much better than caterpillars, but less well than most adult insects.

Some caterpillars have been shown to possess three visual pigments with wavelength sensitivities similar to those of adults. So some larvae, at least, can discriminate between colors.

3.6 Conclusions

It is important to appreciate how the sensory systems of insects work in order to understand behavior. It is also important, however, to appreciate that the insects

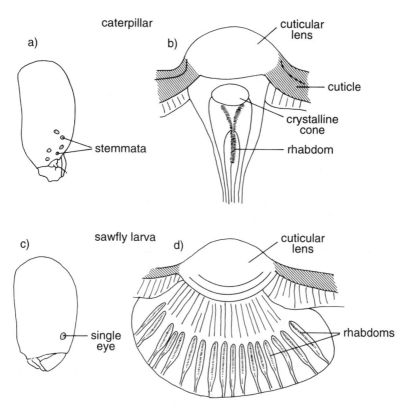

Figure 3.25. Larval eyes. **a,b)** A caterpillar. a) shows the position of the stemmata on the side of the head. b) shows the detail of their strucutre. There ıs only one rhabdom. **c,d)** A sawfly larva. c) shows the position of the single eye on the side of the head. d) shows details of the structure. There are numerous rhabdoms. Consequently, the sawfly larva has much better visual resolution than the caterpillar.

are not automata, responding to a specific stimulus with a set response. A host-specific odor, for example, only elicits a behavioral response under certain circumstances. The response depends on integration of information about the odor with all the other factors operating on and in the insect. If it is the wrong time of day, or the insect is fully satiated, or the wind speed is too high, or some other factor is inappropriate, the insect will not respond. Consequently, the behavior we observe will often not be that which we might have expected, and a proper appreciation can only be obtained with a knowledge of the whole organism.

Further Reading

Dethier, V.G. 1976. *The Hungry Fly*. Harvard.

Frazier, J.L. 1992. How animals perceive secondary plant compounds. *In* Rosenthal,

G.A. and Berenbaum, M.R. *Herbivores. Their Interactions with Secondary Plant Metabolites*, vol. 2. Academic Press, San Diego, pp. 89–134.

References (* indicates review)

Taste

Blaney, W.M. and Simmonds, M.S.J. 1988. Food selection in adults and larvae of three species of Lepidoptera: a behavioral and electrophysiological study. Entomologia exp.appl. 49: 111–121.

Blaney, W.M. and Simmonds, M.S.J. 1990. A behavioural and electrophysiological study of the role of tarsal chemoreceptors in feeding by adults of *Spodoptera*, *Heliothis virescens* and *Helicoverpa armigera*. J.Insect Physiol. 36: 743–756.

*Städler, E. 1984. Contact chemoreception. *In* Bell, W.J. and Cardé, R.T. (eds.), *Chemical Ecology of Insects*. Chapman and Hall, New York, pp. 3–35.

Distribution and abundance of contact chemoreceptors

*Chapman, R.F. 1982. Chemoreception: the significance of receptor numbers. Adv.Insect Physiol. 16: 247–356.

Chapman, R.F. and Fraser, J. 1989. The chemosensory system of the monophagous grasshopper, *Bootettix argentatus* Bruner (Orthoptera: Acrididae). Int.J.Insect Morphol. & Embryol. 18: 111–118.

Variability of response

Abisgold, J.D. and Simpson, S.J. 1988. The effect of dietary protein levels and haemolymph composition on the sensitivity of the maxillary palp chemoreceptors in locusts. J.Exp.Biol. 135: 215–229.

Bernays, E.A., Blaney, W.M. and Chapman, R.F. 1972. Changes in chemoreceptor sensilla on the maxillary palps of *Locusta migratoria* in relation to feeding. J.Exp.Biol. 57: 745–753.

Schoonhoven, L.M., Simmonds, M.S.J. and Blaney, W.M. 1991. Changes in the responsiveness of the maxillary styloconic sensilla of *Spodoptera littoralis* to inositol and sinigrin correlate with feeding behaviour during the final larval stadium. J.Insect Physiol. 37: 261–268.

Simmonds, M.S.J., Schoonhoven, L.M. and Blaney, W.M. 1991. Daily changes in the responsiveness of taste receptors correlate with feeding behaviour in larvae of *Spodoptera littoralis*. Entomologia Exp.Appl. 61: 73–81.

Simmonds, M.S.J., Simpson, S.J. and Blaney, W.M. 1992. Dietary selection behavior in *Spodoptera littoralis*: the effect of conditioning diet and conditioning period on neural responsiveness and selection behaviour. J.Exp.Biol. 162: 73–90.

Simpson, S.J., James, S., Simmonds, M.S.J. and Blaney, W.M. 1991. Variation in chemosensitivity and the control of dietary selection behaviour in the locust. Appetite 17: 141–154.

Sensory coding

Chapman, R.F., Ascoli-Christensen, A. and White, P.R. 1991. Sensory coding for feeding deterrence in the grasshopper *Schistocerca americana*. J.Exp.Biol. 158: 241–259.

Dethier, V.G. and Crnjar, R.M. 1982. Candidate codes in the gustatory system of caterpillars. J.Gen.Physiol. 79: 549–569.

*Schoonhoven, L.M 1987. What makes a caterpillar eat? The sensory code underlying feeding behavior. *In* Chapman, R.F., Bernays, E.A. and Stoffolano, J.G. (eds.) *Perspectives in Chemoreception and Behavior*. Springer-Verlag, New York, pp. 69–97.

Schoonhoven, L.M. and Blom, F. 1988. Chemoreception and feeding behaviour in a caterpillar: towards a model of brain functioning in insects. Entomologia Exp.Appl. 49: 123–129.

The sense of smell

Guerin, P.M. and Städler, E. 1982. Host odour perception in three phytophagous Diptera—a comparative study. *In* Visser, J.H. and Minks, A.K. (eds.) *Insect-Plant Relationships*. Centre for Agricultural Publishing, Wageningen, pp. 95–105.

Guerin, P.M. and Visser, J.H. 1980. Electroantennogram responses of the carrot fly, *Psila rosae*, to volatile plant components. Physiol.Entomol. 5: 111–119.

Guerin, P.M., Städler, E. and Buser, H.R. 1983. Identification of host plant attractants for the carrot fly, *Psila rosae*. J.Chem.Ecol. 9: 843–861.

Kafka, W.A. 1971. Specificity of odor-molecule interaction in single cells. *In* Okloff, G. and Thomas, A.F. (eds.) *Gustation and Olfaction*. Academic Press, London.

Light, D.M. and Jang, E.B. 1987. Electroantennogram responses of the oriental fruit fly, *Dacus dorsalis*, to a spectrum of alcohol and aldehyde plant volatiles. Entomologia Exp.Appl. 45: 55–64.

Ma, W.-C. and Visser, J.H. 1978. Single unit analysis of odour quality coding by the olfactory antennal system of the Colorado beetle. Entomologia Exp.Appl. 24: 520–533.

*Mustaparta, H. 1984. Olfaction. *In* Bell, W.J. and Cardé, R.T. (eds.) *Chemical Ecology of Insects*. Chapman and Hall, New York, pp. 37–70.

*Visser, J.H. 1986. Host odor perception in phytophagous insects. A.Rev.Entomol. 31: 121–144.

White, P.R. and Chapman, R.F. 1990. Olfactory sensitivity of gomphocerine grasshoppers to the odours of host and non-host plants. Entomologia Exp.Appl. 55: 205–212.

*Zacharuk, R.Y. 1985. Antennae and sensilla. *In* Kerkut, G.A. and Gilbert, L.I. (eds.) *Comprehensive Insect Physiology, Biochemistry and Pharmacology* vol. 6. Pergamon Press, Oxford, pp. 1–69.

Distribution and abundance of olfactory sensilla

Greenwood, M. and Chapman, R.F. 1984. Differences in numbers of sensilla on the antennae of solitarious and gregarious *Locusta migratoria* L. (Orthoptera: Acrididae). Int.J.Insect Morphol. & Embryol. 13: 295–301.

Integration of olfactory inputs

*Hildebrand, J.G. and Montague, R.A. 1985. Functional organization of olfactory pathways in the central nervous system of *Manduca sexta*. *In* Payne, T.L., Birch, M.C. and Kennedy, C.E.J. (eds.) *Mechanisms in Insect Olfaction*. Clarendon Press, Oxford, pp. 279–285.

The sense of touch

*McIver, S.B. 1985. Mechanoreception. *In* Kerkut, G.A. and Gilbert, L.I. (eds.) *Comprehensive Insect Physiology, Biochemistry and Pharmacology* vol. 6. Pergamon Press, Oxford, pp. 71–132.

Zacharuk, R.Y. and Albert, P.J. 1978. Ultrastructure and function of scolopophorous sensilla in the mandible of an elaterid larva (Coleoptera). Can.J.Zool. 56: 246–259.

Vision

*Menzel, R. and Backhaus, W. 1991. Colour vision in insects. *In* Gouras, P. (ed.) *The Perception of Colour*. CRC Press, Boca Raton, pp. 262–293.

*Trujillo-Cenoz, O. 1985. The eye: development, structure and neural connections. *In* Kerkut, G.A. and Gilbert, L.I. (eds.) *Comprehensive Insect Physiology, Biochemistry and Pharmacology* vol. 6. Pergamon Press, Oxford, pp. 171–223.

*White, R.H. 1985. Insect visual pigments and color vision. *In* Kerkut, G.A. and Gilbert, L.I. (eds.) *Comprehensive Insect Physiology, Biochemistry and Pharmacology* vol. 6. Pergamon Press, Oxford, pp. 431–493.

4

Behavior: The Process of Host-Plant Selection

As we have seen in Chapter 1, all phytophagous insects exhibit some degree of selectivity in the foods they eat. Consequently, they will all be faced with the necessity of selecting an appropriate host at some stage in their life history. In this chapter we describe the more important behavior patterns associated with selection.

Many adult holometabolous insects (flies, beetles, and butterflies and moths) lay their eggs on an appropriate host plant for larval development, and the larva may appear never to be faced with having to find a host. But critical studies of larval behavior reveal that this is not correct. Females do not always select the most appropriate host and newly hatched larvae may reject the plant on which they hatch. If they are to survive, they must find, and recognize, a better host plant. Other species lay their eggs away from the larval host plant, and the larvae, on hatching, must find a host. Even where the female choice has been appropriate, larvae often move to new plants as their food supply is diminished, and all are able to determine the quality of different parts within their host. So we should expect that larvae of even the most host-specific insects have the capacity to distinguish host from nonhost plant, and to determine the quality of the host. Experimental studies show that this is the case.

Host-plant selection involves not only choosing the right species of plant, but also selecting an individual plant within that species that is, or will be, suitable for feeding, survival and development. In recent years the attention of biologists has focused almost exclusively on the mechanisms of finding the appropriate host species, and the process of choosing plants of appropriate quality has been largely neglected. The importance of selection within the species is clearly indicated by field studies on butterfly oviposition where it is common to see a taxonomically appropriate host rejected (see Section 4.3). Although descriptions of feeding insects rejecting an appropriate host species are uncommon, it is sometimes observed in the laboratory that an insect will reject a favored plant and then, almost immediately feed on another plant of the same species, or even some other part of the same plant. The importance of intraspecific variation in host plant acceptability is recognized by plant breeders when they produce insect-resistant cultivars, although, of course, factors other than behavior are involved in varietal resistance.

Because of the tendency to focus on the taxonomic aspects of host selection, there is little information about the factors used by insects in selecting for plant quality within the host range. What information there is, suggests that the behavioral mechanisms involved are not fundamentally different in the two aspects of host-plant selection. For this reason we have not attempted to separate them in the following account, but it will be obvious where we discuss aspects of choosing within the plant species.

Host-plant selection poses two problems for the insect. First, it must be able to detect and locate its host from a distance. Second, having arrived at the plant, it must confirm the appropriateness of that plant in terms of its species and its quality.

Because of their importance in host-plant selection, the chemicals produced by plants are classified according to their effects on the behavior of insects. The terms used in this classification were clearly defined by Dethier, Barton Browne and Smith in 1960. The most important definitions are:

- **attractant:** a chemical that causes an insect to make oriented movements towards the source of the stimulus
- **repellent:** a chemical that causes an insect to make oriented movements away from the source
- **feeding or oviposition stimulant:** a chemical that elicits feeding or oviposition. "Feeding stimulant" is synonymous with "phagostimulant"
- **deterrent:** a chemical that inhibits feeding or oviposition.

Notice that attractant and repellent have an orientation component and they can be effective at some distance from the plant. By contrast, phagostimulation and deterrence only occur when the insect is in a position to touch or bite the plant. They have no orientation component so that, for example, a deterrent compound may cause an insect to stop feeding, but it does not cause the insect to move away from the food.

This is a functional classification based on each insect's biology. A compound that is deterrent to one species may be phagostimulatory to another. Whatever the chemical, its effect will be concentration dependent. Below a certain threshold concentration the insect will be unable to detect it. Above the threshold the effect usually increases as the concentration is increased.

In order to understand the effects of particular compounds it is necessary to examine them individually. However, in the plant the chemicals are present in a complex mixture. Since chemicals often interact with each other at the sensory receptors (see Section 3.2.2), and effects within the central nervous system are not necessarily additive, it cannot be assumed that the behavioral response of an insect to a mixture of chemicals will be the sum of its responses to the individual components.

4.1 Host finding

Attraction from a distance may involve smell or vision, or both. Plant odors can be taxon-specific and the insect's olfactory system often has the capacity to distinguish these odors from others. Plant shape and color, on the other hand, are usually less characteristic because they are variable even within the species. Visually mediated responses are, therefore, relatively unspecific, but when made in the context of a specific odor, play a key role in host location.

4.1.1 Odor-induced attraction

There are many examples of insects being attracted to the odors of their host plants, both by flying (Table 4.1) and by walking or crawling (Table 4.2). Most examples concern monophagous or oligophagous species, but in a few cases there is evidence that polyphagous species are attracted to their hosts. For example, the moths *Heliothis virescens* and *Trichoplusia ni*, both of which are polyphagous, are attracted by the odors of a variety of different plants in wind tunnels. Nymphs of the locust, *Schistocerca gregaria*, move upwind towards the odor of grass in a wind tunnel, although this only occurs when the insects have been deprived of food for some time. Root-feeding larvae, like the larvae of the beetle, *Diabrotica*, crawl through the soil towards grass roots. It is evident in all these cases that odors are involved, but how are the insects able to orient and move to the odor source?

Because of air turbulence, concentration gradients of plant odors are unlikely to exist more than a few centimeters away from a plant. Instead, the air coming from an odor source contains pockets of odor-carrying air, carried in a mass of clean, nonodorous air. These pockets of odor are carried downwind from the plant so that an insect some distance from the source will perceive a series of bursts of odor separated by periods without odor. The concentration of odor within a pocket, or burst, is very variable and, even some distance downwind of the source, some pockets will still carry high concentrations (Fig. 4.1). The durations of bursts and the intervals between them tend to increase at greater distances from the source. Consequently, although on average the odor concentration measured over a period will be lower at greater distances from the source, the insect has no immediate way of determining the direction in which it must move to reach the source by monitoring the characteristics of the bursts of odor it perceives. In other words, at distances beyond a few centimeters from a plant, the insect cannot follow an odor gradient to its source because there is no gradient to follow.

Table 4.1. *Some insects that are known to fly towards the sources of odors of host plants or host-related chemicals. Entries are based on catches in field traps or laboratory observations in a wind tunnel. In most of these cases, the behavior is related to oviposition. Note: "host-specific compounds" means that the compounds are characteristic of the host; "host-related compounds" means compounds produced by the host plant, but also by other plants that are not hosts.*

Insect	Common name	Host plant	Diet breadth	Odor source
Coleoptera				
Diabrotica spp.	corn rootworms	many	polyphagous	host-related compounds
Glischrochilus spp.	nitidulid beetles	many	polyphagous	mixture of host-related compounds (e.g., acetaldehyde, ethanol)
Phyllotreta spp.	flea beetles	brassicas	oligophagous	host-specific compounds (e.g., allylisothiocyanate)
Smicronyx fulvus	red sunflower seed weevil	sunflower	oligophagous	mixture of host-related compounds (see table 4.3)
Diptera				
Dacus tryoni	Queensland fruit fly	fruits	polyphagous	host extracts; fruit-related chemicals (e.g., ethyl butyrate)
Delia radicum	cabbage root fly	brassicas	oligophagous	host-specific compounds (e.g., allylisothiocyanate)
Delia antiqua	onion fly	onion	oligophagous	host-specific compounds (e.g., dipropyl disulfide)
Psila rosae	carrot root fly	carrot	monophagous	host-specific compounds (e.g., *trans*-asarone)
Rhagoletis pomonella	apple maggot fly	apple	oligophagous	host extracts; mixture of host-related compounds (e.g., propyl hexanoate, hexyl propionate)
Lepidoptera				
Acrolepiopsis assectella	leek moth	leek	monophagous	host-specific compounds (e.g., propylthiosulfinate)
Agraulis vanillae	gulf fritillary	*Passiflora*	monophagous	host plant
Amyelois transitella	naval orangeworm	almonds	oligophagous	host-related compounds
*Ctenucha virginica	—	*Eupatorium*	—	host-specific compounds (e.g., dihydroxydanaidal)
Heliothis subflexa	—	ground cherry	monophagous	host plant extracts
Heliothis virenscens	tobacco budworm	many	polyphagous	host plant extracts and volatiles
Manduca sexta	tobacco hornworm	Solanaceae	oligophagous	host plant
Trichoplusia ni	cabbage looper	many	polyphagous	host plants

*an example of pharmacophagy (Section 4.2.4.2).

98

Table 4.2. Some insects that are known to walk or crawl towards the sources of odors of host plants. Entries are based on laboratory studies using a variety of wind tunnels and olfactometers.

Insect	Common name	Host plant	Diet breadth	Odor source
Orthoptera				
Schistocerca gregaria (larva)	desert locust	many	polyphagous	grass
Hemiptera				
Cryptomyzus korschelti	aphid	*Stachys*	oligophagous	host plant
Coleoptera				
Costelytra zealandica (larva)	grass grub	many	polyphagous	host plant roots
Cylas fornicarius	sweet potato weevil	sweet potato	oligophagous	host plant
Diabrotica virgifera (larva)	western corn rootworm	grasses	oligophagous	host plant; host-related compounds[1]
Leptinotarsa decemlineata	Colorado potato beetle	Solanaceae	oligophagous	host plant; host-related compounds[1]
Trirhabda canadensis	goldenrod beetle	goldenrod	oligophagous	host plant
Diptera				
Delia radicum (larva)	cabbage rootfly	brassicas	oligophagous	host related and host-specific compounds[1]
Psila rosae (larva)	carrot fly	carrot	monophagous	host-specific compounds[1]
Lepidoptera				
Papilio demoleus (larva)	citrus swallowtail	citrus	oligophagous	host-related compounds[1]

[1]See caption to Table 4.1.

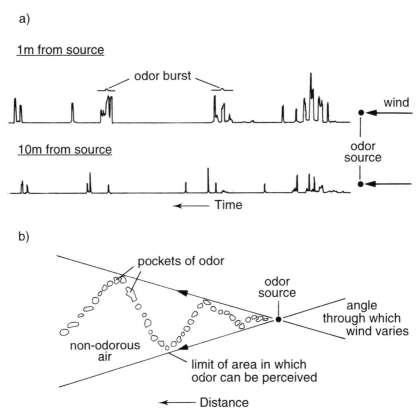

Figure 4.1. Dispersal of odor from a source. **a)** A stationary insect will perceive pulses of odor separated by intervals of clean air. At greater distances from the source, the pattern is basically the same. Although, when averaged over time, the concentration is lower at 10m than at 1m, some odor pulses will carry high concentrations. Odor concentration is indicated by the height of the deflection above the baseline. **b)** As the wind swings about, the odor is broken into a series of pockets (equivalent to the bursts in (a)). Insects will only perceive the odor if they are within the limits of the meandering plume. They must move upwind to reach the source.

How then can odor be involved in host finding? There are two stages: first, arousal and then, orientation. Arousal of an insect by an odor puts it into a state of readiness to respond to some further stimulus. In some adult insects, arousal will often lead to take-off. Orientation occurs after take-off. In other cases, odor-induced arousal causes the insect to orient while still on the ground. This is true of the adults of some species and also of nonflying forms. Orientation with respect to wind is called an anemotaxis, and upwind orientation is positive anemotaxis. Orientation to wind that results from stimulation by an odor is called odor-induced anemotaxis.

Several species of insect have been shown in the laboratory to respond to odor by walking upwind (Table 4.2), presumably using the antennae or mechanoreceptor hairs on the head to determine the direction of wind movement. Larvae of the desert locust, *Schistocerca gregaria*, and adult Colorado potato beetles, *Leptinotarsa decemlineata*, are examples. Wingless females of the aphid, *Cryptomyzus korschelti*, have also been shown to walk upwind in the odor of their host plant in a wind tunnel. Although there are few reports of upwind walking under natural conditions, it has been demonstrated that goldenrod leaf beetles, *Trirhabda canadensis*, walked towards their hosts from four meters away. This was not a purely visual response because the beetles were not attracted by a target of nonhost plants of similar size and with similar leaf shapes.

Some insects that fly towards an odor source also appear to orient to the wind using mechanical stimuli while they are still on the ground. The cabbage root fly, *Delia radicum*, and the onion fly, *Delia antiqua*, are day-flying insects, rather smaller than a house fly. When stimulated by a host odor, these flies orient into the wind and then take off. They make only short flights of 50 to 150 cm before landing, reorienting and taking off again. This series of short flights close to the ground will eventually bring them close to the odor source. If the odor is lost during one of the flights, the fly remains on the ground until it is restimulated or takes off in some direction unrelated to the odor source (Fig. 4.2a). In the cabbage root fly only gravid females, those with fully developed eggs, respond to the odor, but both sexes of the onion fly do so.

Insects making more sustained flights, perhaps from greater distances, cannot use the same method for detecting wind direction. Once airborne, any organism is carried along by the moving air mass of which it now forms a part. Although an insect can monitor its own movement through the air using wind sensitive hairs and antennae, there are no mechanosensory means by which it can tell the direction of movement of the airmass which it is in. It can do so visually, however. An insect flying forward in still air will perceive images moving over the eye from front to back at a moderate speed. If it is flying into a wind, it will continue to make headway provided it is flying through the air faster than the wind is carrying it backwards, or, in other words, its airspeed exceeds the speed of the wind. Many insects are able to increase their airspeeds so as to maintain a more or less constant speed over the ground (groundspeed), as determined by the rate of image movement across the eye. Ultimately, however, if the windspeed is too high, the insect will be unable to compensate and it will be blown backwards. In this case, the movement of images across the eye will be from back to front. Insects will not tolerate this, and usually land or turn to fly with the wind.

The use of visual images by flying or swimming insects to maintain orientation to a current flow is called an optomotor reaction. It enables the insect to maintain an orientation at any angle to the wind, not just directly up- or downwind. If the insect is unable to see the pattern of objects on the ground it cannot orient.

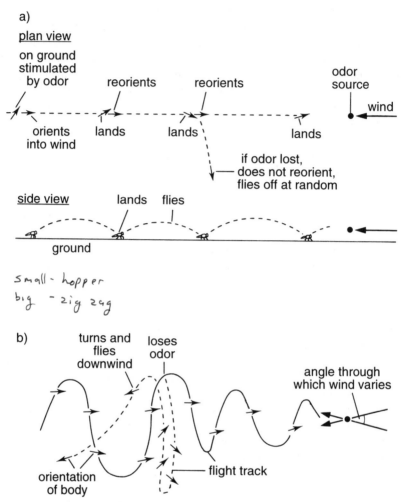

Figure 4.2. Orientation to odor. **a)** A small insect on the ground orients into wind using mechanoreceptors when stimulated by odor from a host plant. It takes off and flies a short distance before landing. If it continues to perceive odor, it reorients and takes off again. It progresses upwind in a series of short flights separated by odor-induced reorientation to the wind. If it loses the odor, it does not reorient, but flies off in any direction. **b)** A large insect in sustained flight orients to wind carrying odor from a host plant by an odor-modulated optomotor reaction. As long as it continues to be stimulated, it zig-zags upwind. The turns are not induced by reaching the edge of the odor plume; they are the result of a program in the central nervous system. If the insect loses contact with the odor, it reduces its airspeed and swings from side to side without making progress towards the odor source. If contact with the odor is not regained, it turns and flies downwind.

In night-flying insects the eyes are adapted to maximize the use of the available light at the expense of acuity (see Section 3.5).

The use of an optomotor response to locate host plants is probably widespread amongst phytophagous insects. Females of *Manduca sexta* use an optomotor response as they zigzag upwind towards a host plant in a wind tunnel. The response is the same as that used by males when they move towards a female emitting pheromone, or to an artificial pheromone source. This pattern of movement is programmed into the insect's nervous system and continues as long as the insect is stimulated by odor. If it loses contact with the odor, the insect stops moving upwind, directing its flight more across the wind line. If it fails to remake contact with the odor, it ceases to make progress upwind and then turns to fly downwind (Fig. 4.2b).

How very small insects, such as winged aphids, can locate their hosts is still unclear. The problem here is that the insects are weak fliers and maximum airspeeds will rarely exceed 1.5 m s^{-1}. Consequently, they will be unable to make headway against any but the lightest winds. Yet any insect reacting to and approaching an odor source must do so from downwind of the object.

The consensus arising from earlier work is that polyphagous aphids, like *Aphis fabae* and *Myzus persicae*, arrive at plants in an undirected manner. The process is not quite random because the insects favor landing on certain colors, but odor apparently does not play a part. Having landed on a plant, the insect remains if it is an appropriate host, as determined by odor and other characteristics; if it is not, the insect takes off again.

However, there is an increasing number of examples of small insects that do make oriented flights towards a host. Most of these concern parasitoids, but some also relate to phytophagous insects. In one example, carrot aphids, *Cavariella aegopodii*, were caught in water traps baited with carvone, an isomer of one of the components of host odor. Most of the aphids were caught during periods when the windspeed was less than 1 m s^{-1}, and they appeared to be flying upwind towards the odor source from more than 1 m away. They may have been orienting anemotactically, but were probably also responding to the traps in the presence of odor as visual targets to which they could direct their flights. Traps without the odor of carvone were ineffective.

What we do not know in this and similar instances is how the insects came to be close enough to the traps to reach them during the brief periods when the wind speed was low enough for them to approach the traps. Had they been carried for a long distance downwind and, by the process of 'random' fallout, landed close to the traps? Or had they flown upwind for some distance, stimulated by the odor, landing whenever the wind was too strong for them to make headway? These are critical unanswered questions in our understanding of odor-induced attraction to host plants by phytophagous insects.

It is probably true that many plant species do not produce an odor based on some characteristic chemical, but all plants do produce a range of volatile compounds as a result of the oxidative degradation of leaf lipids. This process gives rise to hexanal and hexanol and, from these, a series of other compounds with chain lengths of six carbon atoms is produced. The compounds are called the "green leaf volatiles" (see Section 2.1). They differ from plant to plant to some extent, but the same range of compounds is common to many species of plants. Some insects have been shown to be attracted by single "green leaf chemicals," but this, of course, cannot result in anything more than a very general attraction to vegetation.

However, some plants produce a mixture of the green leaf volatiles that is characteristic of the species. Experiments in a wind tunnel strongly suggest that the Colorado potato beetle, *Leptinotarsa decemlineata*, is attracted by the specific combination of green leaf volatiles produced by potato and not by any other combination of odors. The chemicals that are important are *trans*-2-hexenal, *cis*-3-hexanyl acetate, *cis*-3-hexenol and *trans*-2-hexenol.

In the case of some oligophagous insects, attraction is induced by the odors of compounds that are specific to the host plant. For example, crucifer-feeding insects are attracted by isothiocyanates derived from the glucosinolates that characterize the plants chemically. Onion flies are attracted by volatile sulfur-containing compounds, such as dipropyldisulfide, that characterize the onion (see Table 4.1).

It may be that under natural conditions a characteristic combination of volatiles is important. In some species single chemicals fail to induce any attraction; only a mixture is effective. This is illustrated by a study of the red sunflower seed weevil, *Smicronyx fulvus*. Effective attraction is only produced by a mixture of five terpenoids in the ratio in which they occur in the air above susceptible sunflower cultivars. The individual chemicals, or combinations of two or three are almost completely ineffective. So is the mixture if the components are in different proportions (Table 4.3).

Under natural conditions, the odor of one type of plant will commonly be mixed with the odors of other species growing in the same habitat. This may affect the responses of an insect to its host odor, and this has been clearly demonstrated in laboratory studies with the Colorado potato beetle. This insect is attracted by the odor of potato, but, if the host odor is mixed with the odors of tomato or cabbage, attraction no longer occurs. In another laboratory study, on the black swallowtail butterfly, *Papilio polyxenes*, it has been shown that females ready to lay eggs are stimulated to land on artificial leaves if these are treated with volatiles from carrot leaves. Carrot is a host plant for this species. But the response is inhibited if volatiles from cabbage are also present.

This principle of masking the odor of one plant with that of another is utilized in the agricultural practice of intercropping, where two or more different crops are planted together in the same field. For example, infestation of cabbage by the flea beetle, *Phyllotreta crucifera*, is reduced if the cabbage is intercropped with tomato. The odor of tomato is believed to repel the insects, though this has

Marigolds

Table 4.3. *Attraction of the red sunflower seed weevil, Smicronyx fulvus, to traps baited with different components, in different proportions, of sunflower odor. Numbers caught are expressed as a percentage of the number in the complete blend (= 100%). The individual chemicals are produced by many plants. Only the combination of all five in the correct ratio is characteristic of the host. + = present in mixture, — = absent from mixture (after Roseland et al., 1992).*

	Mixture					Number Caught/ Trap (%)
	α-Pinene	β-Pinene	Camphene	Limonene	Bornyl acetate	
Control	—	—	—	—	—	16
Complete—Correct Ratio	+	+	+	+	+	100
Complete—Incorrect Ratio	+	+	+	+	+	21
Incomplete	—	+	—	+	+	31
Incomplete	+	—	+	+	—	16
Incomplete	+	+	—	—	+	4

not been demonstrated unequivocally. In this case, the mechanism might be different from that occurring in the Colorado potato beetle. The potato beetle is attracted by a specific mixture of green leaf volatiles, so the wrong mixture, resulting from the presence of another plant, will not provide an appropriate signal. *P.crucifera*, however, is attracted by specific chemical components of the cabbage and probably has olfactory receptors responding specifically to these compounds. The response of these receptors is less likely to be affected by chemicals from other plants. It is possible that the odor of the nonhost stimulates other receptors and the input from these produces a change in the insect's behavior.

There is very little critical information concerning the distances from which insects are able to perceive the odor of a host plant. Obviously this distance will vary greatly, depending on the rate of release of volatiles from the plant, whether the plant is an isolated individual or growing in a stand of other plants of the same or different species, the nature of the terrain, and wind conditions. One study has shown that the onion fly, *Delia antiqua*, orients and takes off into an odor-laden wind 100 m from the odor source. In this case, the odor was dipropyldisulfide, one of the characteristic chemicals of onion, and the experiment was carried out in fallow, bare-soil fields.

Once an insect is close to a host plant, the plant odor may also stimulate landing. In *Papilio polyxenes* the number of landings made by females and the number of eggs laid on artificial leaves was greatly enhanced by the presence of host (carrot) volatiles. This occurred even when the odor did not come directly from the leaves, but pervaded the whole cage in which the experiments were carried out. This shows that the landing response was not directed by a gradient of odor, since no gradient existed around the leaves. Probably the odor caused the insects to respond visually to the leaves. As with odor-induced attraction to a plant from a distance, the odor has an arousal effect, making the insect responsive to the sight of the leaves.

Movement towards the source along an odor gradient is only possible when an insect is very close to a plant and the gradient is very steep. This has been demonstrated, in particular, in a number of insect larvae feeding on roots. In the soil, air movements are minimal so that odor gradients are much more likely to persist than they are above the ground. Five compounds, isolated from the vapor surrounding whole carrots, were effective attractants for the larvae of the carrot fly, *Psila rosae*. The compounds were (-)-bornyl acetate, 2,4-dimethyl styrene, biphenyl, α-ionone and β-ionone. These compounds, collectively, only comprised 0.6% of the total volatiles present in the carrot vapor. Other compounds, present in greater quantity, had no behavioral effect. One, *trans*-2-nonenal, was consistently repellent. In general, root-feeding larvae respond to odor gradients of host chemicals, as in the example above, as well as to more widely occurring plant compounds, including carbon dioxide. In these experiments the larvae responded when they were 2.5 cm from the odor source. In other experiments with insect larvae, movement along an odor gradient has always been demonstrated over distances of only a few centimeters.

In a number of experiments using a T-tunnel or a wire grid with no air movement, caterpillars have been shown to turn and move towards, or away from, the sources of plant odor. These responses indicate that the insect was able to perceive a gradient of odor by the differential stimulation of olfactory receptors on the two sides of the head.

4.1.2 Visual attraction

Visual attraction might result from responding to the color or the form of the host plant. Because these vary so greatly within a plant species, visual responses often only occur with an appropriate olfactory stimulus. It is obvious, however, that vision will often be of critical importance in the final stages of host finding, partly because the lack of odor gradients makes it most unlikely that odor would lead an insect directly to a plant, but also because a flying insect necessarily uses vision during landing.

4.1.2.1 Shape and size

Several phytophagous insects have been shown to move towards objects or patterns placed in their visual field. Usually these objects consist of a dark pattern against a light background and the insect is attracted by the contrast. Larvae of the desert locust, *Schistocerca gregaria*, for example, have been shown to move towards objects rising above the vegetation, and many tropical African grasshoppers move into taller vegetation, or even trees, at night. This behavior is not necessarily related to host finding, but indicates the ability of the insects to perceive and respond to visual targets.

At closer quarters, grasshoppers move towards and climb up narrow vertical targets in preference to broader targets or to those at an angle to the vertical. If the target is large, the insect moves towards an edge where there is maximum visual contrast (Fig. 4.3).

The "size" of a target is determined by the angle that it subtends at the eye. So a small object close up will appear the same as a large object farther away (some insects do have the ability to determine their distance from an object, and in these cases this will be an oversimplification). This has been most fully studied in the caterpillar of the citrus swallowtail, *Papilio demoleus*. These insects were shown to turn towards an object provided the vertical or horizontal angles subtended at the eyes exceeded 20°. Narrow vertical targets were no more attractive than wide, low targets. This contrasts with the results obtained with grasshoppers and probably reflects the very limited visual acuity of the caterpillars (see Section 3.5). Nearly all larvae responded when both vertical and horizontal angles exceeded 50° (Fig. 4.4). The distance of the screen from the insect,

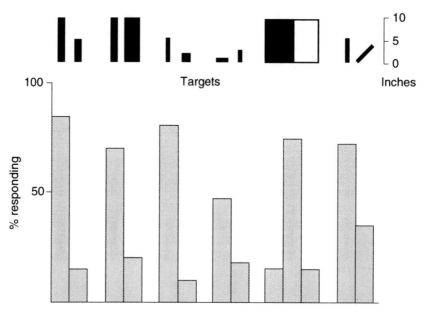

Figure 4.3. Responses of the larvae of *Schistocerca gregaria* to visual targets. Larvae were in a white-walled, circular arena and started in the center, 12 inches from the targets which were black and presented in pairs, as shown. Many insects did not respond to small targets so the percentages do not necessarily add to 100. The scale shows the sizes of the targets (after Wallace, 1958).

between 2 and 8 cm, was not important provided the apparent size remained the same. In practice, this means that the closer a caterpillar is to an object the more likely it is to turn towards it because the subtended angles will increase as it gets closer.

For ovipositing butterflies, leaf shape plays a key role in the process of discriminating the host from other plants. Examples of this are given in section 4.3. The response to leaf shape often includes an element of learning (see Sections 4.3 and 6.4).

Shape may interact with color, as has been demonstrated in the apple maggot fly, *Rhagoletis pomonella*. The flies are attracted to yellow rectangles, but not to red, black or white rectangles. On the other hand, yellow spheres attract very few insects, while red and black spheres attract large numbers (Fig. 4.5). These differences are presumed to reflect key elements in the normal behavior of the fly. Yellow is a component of foliage color (see below) and is possibly involved in tree finding by the insects, while the spheres might represent fruits, the dark color providing increased contrast and making them more easily seen. These flies will attack apples of all shades from green to red.

Experiments with the same species of fly also illustrate the interaction of target

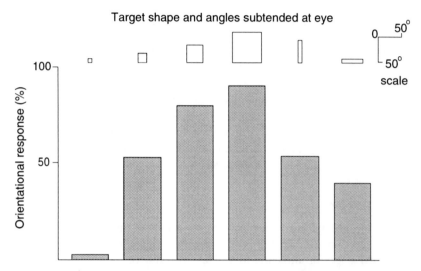

Figure 4.4. Responses of the larvae of *Papilio demoleus* to visual screens subtending different angles at the head of the caterpillar. Larvae were moving on a wire grid and could turn towards, or away from the screen. The orientational response is the net percentage turning toward the screen. In each experiment, the larva was presented with a screen on one side. The angular dimensions of the screen are given at the top of the figure (after Saxena and Khattar, 1977).

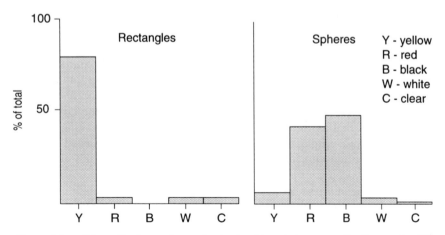

Figure 4.5. Affect of color on form selection by *Rhagoletis pomonella*. Rectangles (30 x 40 cm) and spheres (7.5 cm in diameter) of different colors were suspended close to apple trees and coated with adhesive to trap the flies. The figure shows the percentage of flies caught on each of the traps (after Prokopy, 1968).

shape with odor. An attractant odor increased by three- to fivefold the number of flies landing on rectangles, but had no significant effect on the numbers attracted to spheres.

Learning the size of fruit may also influence the host-finding behavior of some fruit flies. The oriental fruit fly, *Dacus dorsalis*, will oviposit on many different kinds of fruit. In some cases, its ability to discover the fruit depends on its previous experience. If a female is exposed to kumquat fruit for a period, she becomes better at finding this fruit than females who have experienced different fruits. She retains this memory for at least three days even if she is exposed to other fruit during this period. It appears that learning the size of the kumquat is important. On the other hand, experience does not increase the ability of females to find apples.

4.1.2.2 Color

Leaves in general are various shades of green and their spectral reflectance patterns tend to be similar (Fig. 4.6). Nevertheless, leaf color may be important in the final stages of attraction to a plant. In addition, for species that feed or oviposit on flowers, we might expect color to be especially important.

In discussing the responses of insects to color it is important to distinguish between the wavelength of light and its intensity. Wavelength determines which of the visual pigments will absorb the light and so which of the cells in an ommatidium will be stimulated (see Section 3.5). In human terms, this determines the color that we visualize. But colors in nature are never monochromatic, that is, composed of one wavelength only, and the response of an insect, or the color we perceive, is altered by the degree of reflectance of a particular waveband relative to reflectance across the whole spectrum. A monochromatic light is said to be saturated. With the addition of increasing amounts of other wavelengths it becomes progressively more unsaturated. This is illustrated in Fig. 4.7a. Fig 4.7b shows that as the reflectance spectrum from water traps becomes broader, more unsaturated, fewer flower thrips, *Frankliniella occidentalis*, are caught.

The second major factor in color vision is the intensity of the light which, in the case of a plant, refers to the amount of light reflected from the surface. Fig. 4.7c,d shows that as the intensity of light reflected from traps was reduced, fewer thrips were caught, even though the peak wavelengths remained the same. The addition of an ultraviolet component to white light completely suppressed the catch.

Few experiments have been carried out on the behavior of insects in monochromatic light, that is light of a single wavelength, in which the intensity of light was the same for all wavelengths tested. One such study on the large white butterfly, *Pieris brassicae*, examined the effects of wavelength on tarsal drumming and oviposition by the female, as well as on nectar feeding. Oviposition only occurred if the substrate was treated with sinigrin (see below). No drumming was observed at wavelengths below 500 nm or above 600 nm. Maximal drumming

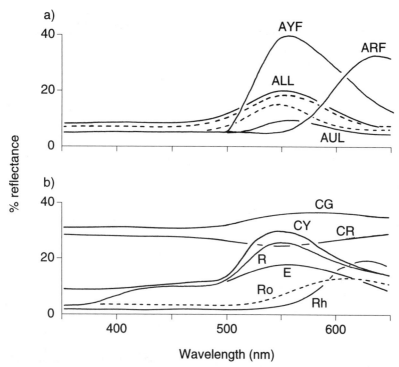

Figure 4.6. Reflectance patterns from various leaves and fruits: **a)** apple and peach leaves and fruits. Reflectances from apple are shown by full lines; reflectances from the upper (below) and lower (above) surfaces of peach leaves by broken lines. Abbreviations: ALL - apple, underside of leaf; AUL - apple, upperside of leaf; AYF - apple, yellow-green fruit; ARF - apple, red fruit. **b)** a range of other plants. For most leaves peak reflectance occurs at about 550nm, but some leaves have a strong red component. Abbreviations: CG - cabbage, green; CR - cabbage, red; CY - cabbage, young; E - *Euphorbia*; R - radish; Rh -*Rhus*; Ro - rose (after Prokopy and Owens, 1978; Prokopy et al., 1983).

and oviposition behavior was observed at wavelengths between 525 and 575 nm. Feeding, on the other hand, occurred primarily when the wavelength was below 500 nm (Fig. 4.8).

Such results have given rise to the concept of wavelength-specific behaviors. The association between wavelength and behavior is presumably programmed into the central nervous system of the insect and it has obvious adaptive value. The butterfly commonly feeds at blue flowers and lays its eggs on green leaves. However, it should not be supposed that the responses of the insect in nature are an inevitable consequence of perceiving certain wavelengths. The insect is not induced to feed by every blue flower that it sees, nor does it drum on every green leaf. In practice, the behavior of the insect is determined by its physiological

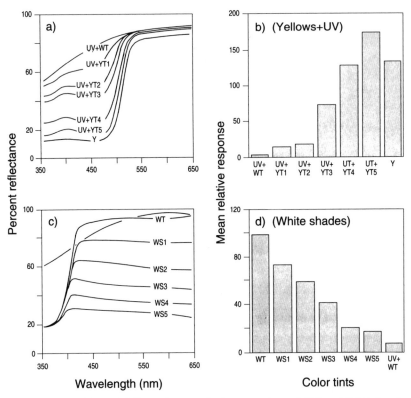

Figure 4.7. Effects of different degrees of color saturation and different light intensities on trap catches of the flower thrips, *Frankliniella occidentalis*. **a,b)** Saturation. a) Shows the percentage of light of different wavelengths reflected from traps colored yellow with different degrees of saturation. b) shows the relative numbers of thrips caught in the traps. The number caught decreased as the degree of unsaturation of the color increased. **c,d)** Intensity. c) shows the percentage of light of different wavelengths reflected from white traps. d) shows the relative number of thrips caught in the traps. The catch decreased as the intensity of reflected light decreased. The trap catch is expressed relative to the number caught in the saturated white traps. The addition of ultraviolet (unlabeled line in c) greatly reduced the catch (after Vernon and Gillespie, 1990).

state and experience. An egg-laying female butterfly can switch from ovipositing to nectar feeding and back again as needs demand. Similarly, during its migratory phase, a winged aphid resting on a yellow leaf shows an increasing tendency to fly up to the blue sky, but after a period in flight has an increasing tendency to land on the yellow again. In addition, a gravid cabbage butterfly that has experienced sinigrin in association with a blue substrate will subsequently land on blue filter paper discs as a potential oviposition substrate in preference to green discs (see below).

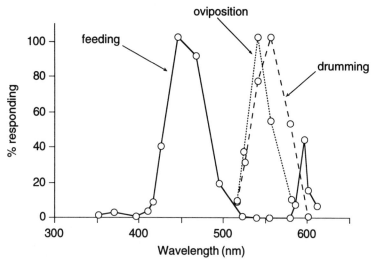

Figure 4.8. Wavelength-specific responses of adult *Pieris brassicae*. The behavior of the insects was observed in monochromatic light at each of the wavelengths indicated by a point. Each type of activity was tested separately. Oviposition only occurred when sinigrin was present in the substrate. Drumming behavior was not examined separately at wavelengths below 500nm, but was never seen during the observations on feeding in this range (after Scherer and Kolb, 1987).

Wavelength-specific behavior must be an evolved adaptation, but not one that dominates the behavior of the insect. We do not know how widespread this phenomenon is. It obviously depends on the insect having a visual system that permits wavelength discrimination.

The importance of leaf reflectance patterns as attractants for insects was established unequivocally in a series of experiments by R.J. Prokopy and his collaborators. They made artificial leaves colored with the same reflectance patterns as real leaves. When given a choice of real leaves of radish, green cabbage and red cabbage, most cabbage root flies, *Delia radicum*, landed on radish. The radish leaves had a peak reflectance around 540nm (Fig. 4.6b). By contrast both of the cabbage varieties had high levels of reflectance over much of the visual range. When these reflectance patterns were mimicked using paints, and eliminating all other possible cues, the flies were strongly attracted to the radish mimic (Fig. 4.9).

It is probably true that color is important in the final stages of attraction of many day-flying phytophagous insects to their hosts. In the field, *Aphis fabae* lands preferentially on yellow leaves and it is common for insects to be caught in yellow traps (Table 4.4). In other insects the preferential response to yellow takes other forms and, for example, *Delia antiqua* lays more eggs when the

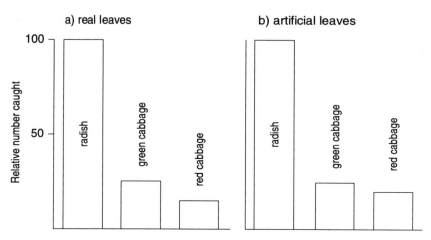

Figure 4.9. The effect of different reflectance patterns on the numbers of cabbage root flies, *Delia radicum*, caught on sticky traps. **a)** The numbers caught on real leaves of three plants. **b)** The numbers caught on artificial leaves with the same reflectance patterns. The relative catches were almost exactly the same, showing that the pattern of reflectance played a key role in attracting the flies. Catches are expressed relative to the numbers on radish (=100). Reflectance patterns of the plants are shown in Fig.4.6. (after Prokopy et al., 1983).

pseudo-onion leaves they are presented with are yellow, rather than any other color. The color affects both the number of runs a female makes on a stem and the number of probings she makes in the surrounding sand (Fig. 4.10; see below for details of behavior). It is not certain why so many insects exhibit this strong response to yellow, but the wavelengths of yellow light, in the range 560–580 nm, are not far off the peak of sensitivity of the insect's green-sensitive pigment. This peak is often close to 540 nm (see Fig. 3.23). Prokopy has suggested that this, together with the high reflectance properties of yellows, causes yellow objects to present a supernormal stimulus to the insects. However, not all phytophagous insect species are attracted to yellow; species that feed and reproduce in flowers are often attracted to blue.

Color patterns may also be important in host-plant selection. This is illustrated by the behavior of ovipositing females of *Heliconius*. These butterflies lay their eggs on the leaves of *Passiflora*, but they tend not to oviposit on leaves that already have eggs on them. This is a visual response to the yellow eggs because if the eggs are painted green, matching the color of the leaf, the butterflies do not distinguish between leaves with and without eggs.

A response to color may be coupled with a chemical cue. For example, the cabbage butterfly, *Pieris rapae*, will not oviposit on any substrate in the absence of a glucosinolate such as sinigrin. If sinigrin is added to the substrate the insects will oviposit on blue, yellow, green or white paper, but few eggs are laid on

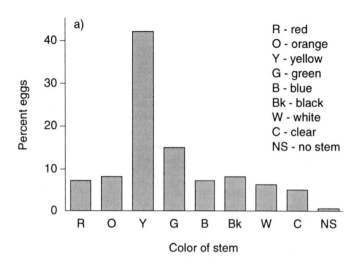

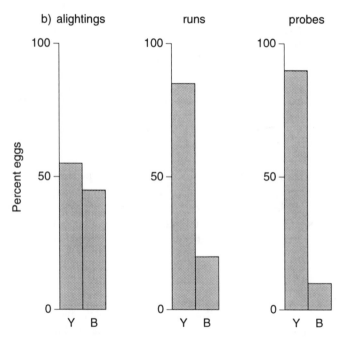

Figure 4.10. Effect of color on oviposition by *Delia antiqua*. **a)** Distribution of eggs in a multiple choice experiment with stem mimics of different colors. **b)** Choice experiments with yellow and blue stem mimics showing how different behaviors contributing to oviposition were affected by color. There was no significant difference in the number of alightings on the two colors, but numbers of stem runs and substrate probes (see text) were much higher on yellow (after Harris and Miller, 1983).

Table 4.4. *Examples of some herbivorous insects that respond positively to yellow in preference to other colors.*

Order	Family	Species	Common name
Orthoptera	Acrididae	*Melanoplus sanguinipes*	migratory grasshopper
Homoptera	Aleyrodidae	*Aleurocanthus woglumi*	citrus black fly
	Aleyrodidae	*Trialeurodes vaporariorum*	greenhouse whitefly
	Aphididae	*Aphis gossypii*	cotton aphid
	Aphididae	*Hyalopterus pruni*	mealy plum aphid
	Aphididae	*Phorodon humuli*	damson-hop aphid
Thysanoptera	Thripidae	*Sericothrips variabilis*	soybean thrips
	Thripidae	*Aeolothrips intermedius*	
	Thripidae	*Thrips pillichi*	
Diptera	Agromyzidae	*Liriomyza sativae*	vegetable leafminer
	Agromyzidae	*Ophiomyia simplex*	asparagus miner
	Anthomyiidae	*Delia antiqua*	onion fly
	Tephritidae	*Anastrepha suspensa*	fruit fly
	Tephritidae	*Ceratitis capitata*	Mediterranean fruit fly
	Tephritidae	*Rhagoletis pomonella*	apple maggot fly
	Otitidae	*Tetanops myopaeformis*	sugarbeet root fly
Coleoptera	Chrysomelidae	*Phaedon cochleariae*	leaf beetle
	Curculionidae	*Anthonomus grandis*	boll weevil
	Curculionidae	*Larinus curtus*	mustard beetle
	Scarabaeidae	*Popillia japonica*	Japanese beetle
Lepidoptera	Noctuidae	*Spodoptera littoralis* (larva)	Egyptian cotton leafworm

red or black. This species can also learn to associate a particular color with the presence of sinigrin. If a gravid female is allowed to touch a green or blue paper impregnated with sinigrin with its forelegs for just 10 seconds, it remembers the color 24 hours later and will land preferentially on filter paper discs of the color it has experienced, even though the paper is only wetted with water and no sinigrin is present (Fig. 4.11).

4.1.3 Conclusions

Attraction to a host plant from a distance often appears to involve both olfactory and visual elements of behavior. The olfactory signal is the indicator of an appropriate host, causing the insect to take off and move towards the source of the odor. But the olfactory system will rarely act alone, and perhaps only does so in those soil dwelling larvae that are directed towards the roots of their host along odor gradients. More usually, a response to odor will be combined with a visual response. The reflectance properties of leaves seem often to determine whether or not a landing is made. Vision also gives the final degree of precision to an approach to a plant enabling the insect to approach and land on the target rather than simply blundering into it.

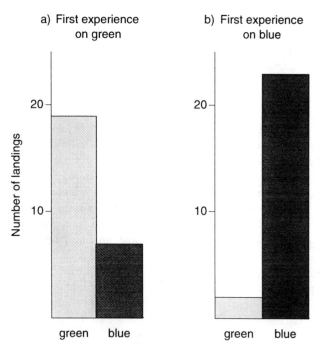

Figure 4.11. Associative learning of colors by *Pieris rapae*. Gravid females were given tarsal contact for 10 seconds with a wet, sinigrin-treated paper which was either green a) or blue b). Twenty-four hours later, the insects were given a choice of wet green and blue papers without any sinigrin. Insects that had experienced green associated with sinigrin showed a strong preference for green; those that had experienced blue with sinigrin preferred the blue (after Traynier, 1984).

In contrast, oviposition-site selection by many butterflies seems to depend largely, and perhaps entirely on visual cues as described in Section 4.3. The extent to which these responses are innate is not clear, and it is certain that accurate discrimination is dependent on a learned association between leaf shape and the chemical characteristics of the leaf perceived when the insect comes into contact with the taxonomically correct host.

4.2 Acceptance of the plant

Once an insect has arrived on a plant it is faced with the decision of whether or not to accept it. Olfaction may still be important, as will contact chemoreception, mechanoreception and vision.

4.2.1 Physical properties of the plant

The importance of the physical properties of the plant, and the range of factors employed in decision-making by the insect, is well illustrated in a study of oviposition behavior by the onion fly, *Delia antiqua*. Having arrived on a plant, or on the soil nearby, the female runs down or up the foliage. After moving vertically on the plant for 4–6 cm she turns round and runs the other way, sometimes also running out onto the surrounding soil. When she has been doing this for about 10 seconds, she extends her proboscis to touch the surface of the plant or the soil, alternately extending and retracting it while she continues to run up and down. This is followed by making a series of superficial probes into soil crevices with her ovipositor. Finally, she pushes her abdomen deep into the soil, stops moving and lays up to five eggs. All this happens quickly. It takes about one minute from the beginning of probing to laying the first egg. After laying one batch of eggs the female may resume running and proboscis extension before laying another batch.

The number of eggs laid by *D. antiqua* over a period is influenced by the color, shape and orientation of the "host." Using model onion leaves, it was shown that a yellow "leaf" induced more oviposition than leaves of other colors (Fig. 4.10); tall narrow "leaves" were more effective than any other shape; and upright "leaves" than those at any other angle (Fig. 4.12). All these characters are, of course, those that most closely resemble a real onion leaf. They only produced their maximum effects when the "plant" was made to smell like an onion by including the odor of dipropyldisulfide.

These differences in egg number were not simply the result of different numbers of flies coming to the targets. The differences shown in Fig. 4.12 arose from decisions made by the flies *after* they had reached the target. It is relatively easy to understand how the fly responds to leaf color and orientation. The cylindrical shape of the target is somehow determined by the fly as it runs over the surface. Perhaps the positions of the legs relative to the body and to each other are involved, as has been suggested for the fly *Lipara lucens* that lays its eggs in shoots of the reed, *Phragmites communis*. During proboscis extension *D. antiqua* is presumably monitoring the chemical qualities of the plant surface.

The importance of physical properties of the leaf in the oviposition behavior of the cabbage root fly, *Delia radicum*, have also been investigated. The leaf surface texture, size and color are important. In addition, these flies tend to follow veins or irregularities in the leaf, or the folds on an artificial leaf. Without these, the movements of the fly are less directed and it is less likely to make the transition from leaf exploration to running on the stem which precedes oviposition.

Smaller features of the plant are also important. The stem borer, *Chilo partellus*, lays its eggs on *Sorghum* plants close to the ground, often on old dry leaves. However, the first instar larvae feed on the young leaves at the top of the plant

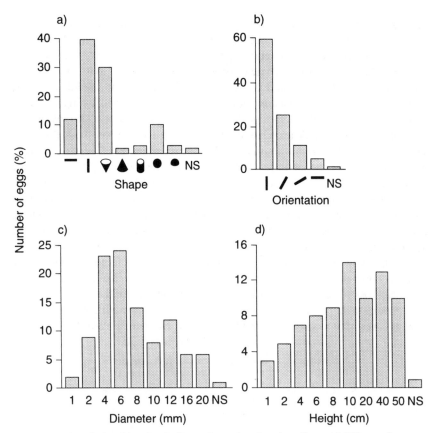

Figure 4.12. Oviposition by *Delia antiqua* showing the effects of different features of the artificial host plant. **a)** Effects of target shape, which is shown along the horizontal axis. **b)** Effects of target orientation, which is shown along the horizontal axis. **c)** Effects of target diameter. **d)** Effects of target height. Targets were presented in multiple choice experiments. Each target stood in an oviposition dish of chopped onion covered by sand. The right-hand column of each figure, labeled NS, shows the percentage of eggs laid in dishes without a target. Most eggs were laid on vertical cylinders 4-6 mm in diameter. Target height had little effect within the range 6-50 cm, but very short targets were less used (after Harris and Miller, 1984).

and their first activity, after hatching, is to find their way to the top of the plant, often one or even two meters above the hatching site. They hatch soon after dawn and move upwards towards the light sky. However, the journey is difficult because the small larvae, less than 2 mm long, often wander on to the leaves which extend obliquely outwards from the culm ("stem") and do not lead to the feeding site. Various factors affect the success of the larvae in making the journey. These include the angles at which leaves project from the culm. Larvae are more likely to fail to complete the climb if the leaves are close to horizontal rather than pointing upwards. The reason for this is that once on a horizontal leaf they have more difficulty returning to the culm than on a more vertical leaf. On some plants the larvae become trapped in hairs in the leaf axil. Other plants have a fold at the leaf base and larvae may become trapped in the folds. So the success of the larvae in reaching the feeding site is affected by many small features. In addition, the nature of the surface wax influences their movement.

Other examples of physical characters preventing feeding are: the inability of early instar *Chorthippus parallelus* to feed on the rolled leaves of *Festuca* because they cannot open the mandibles wide enough to bite the leaf (see Fig. 5.14); and the inability of some small insects to feed because the leaf is too tough.

The leaves of many plants are covered by short hairs, or trichomes, and these may impede insects or even prevent them from feeding on a particular plant or leaf. For example, early instar larvae of the spittle bug, *Philaenus spumaria*, are prevented from feeding on the stem of a host plant if this is hairy, but readily feed on the plant when the hairs are shaved off. The later, larger instars are not affected (Fig. 4.13). The production of hairy cultivars has been used agriculturally to produce resistance against various insects on cotton, wheat and soybeans.

Trichomes also prevent some insect species from ovipositing on plants, but in other cases have a positive value for the insect. *Helicoverpa zea*, for example,

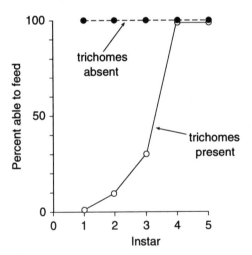

Figure 4.13. Effect of trichomes on the ability of different instars of *Philaenus spumaria* to feed on the stem of *Anaphalis magaritacea* (Asteraceae). Trichomes were shaved from some plants, and, in their absence, all stages of the insect were able to feed (after Hoffman and McEvoy, 1985).

lays more eggs on hairy surfaces at least partly because the ovipositing female is able to hold on to the hairs.

In some plants, glandular trichomes produce a sticky exudate which traps small insects and prevents them from becoming established on the leaf. The best-studied example concerns some species of wild potato, especially *Solanum berthaultii*. The leaves of this species possess two types of trichome: short ones with a four-lobed gland at the tip (type A), and long ones with an ovoid tip (type B) (see Fig. 2.6f). An exudate is continuously discharged from the tips of the type B trichomes. It is viscous so that when an insect lands on the leaf it gets this sticky material on its legs. The exudate also contains sesquiterpenes which excite the insect so that it struggles in the viscous exudate. In doing so, it breaks the glandular heads of the type A trichomes. These contain both phenols and phenoloxidases. When these come together in the air, the phenols are oxidized to a brown substance which immobilizes the insect, so that eventually it dies without having harmed the plant. In some potato varieties, the exudate from the type B trichomes contains β-farnesene, an aphid alarm pheromone.

Some other plants in various families are known to produce exudates that trap small insects. For example, some cultivars of geranium, *Pelargonium*, produce an exudate that traps foxglove aphids, *Acyrthosiphon solani*. If the exudate is washed off, the leaf can be colonized by the aphid.

Many plant species contain laticifers or resin canals within the leaf (see Section 2.4) and when the leaf is damaged these are ruptured. Latex or resin wells out because it is under pressure within the plant. The latex of many plants, including *Asclepias*, contains secondary compounds some of which are feeding deterrents. At least in some plants the concentration of secondary compounds in the latex is very high. For example, the dry weight concentration of cardiac glycosides in the latex of *Asclepias humistrata* averaged 9.5% in one study. This was more than 10 times higher than the levels normally found in whole leaf tissue.

Latexes and resins may also have adverse effects on insects as a result of becoming sticky and hardening when exposed to air. In field experiments with first instar larvae of the monarch butterfly, *Danaus plexippus*, it was found that over 25% of the larvae died as a result of becoming stuck in the latex of *Asclepias humistrata* even though this is a plant on which the insect normally feeds. It is not known if this mortality also involved some degree of poisoning by the high levels of cardiac glycosides present.

Many insects are known to exhibit behavior that circumvents the action of the latex and enables them to feed on latex-containing plants (see Section 4.2.5).

4.2.2 Leaf odor

It is usual to think of plant odors as providing important cues by which insects can recognize their host plants from a distance. But odor may also be important in the final stages of plant acceptance.

The air close to the leaf surface is relatively still. This layer of air is called the boundary layer. Its thickness will depend on the nature of the leaf, whether or not it has a dense layer of trichomes, and on the windspeed. It may commonly be 1–2 mm thick. Odors from the leaf will be present at much higher concentrations in this layer than at greater distances because they will tend not to be carried away by air currents. Small insects, like thrips and aphids, may live much of their lives within the boundary layer. Larger insects may perceive these odors especially during periods of antennation or palpation when they bring their olfactory receptors close to the leaf surface.

Despite the potential importance of the volatiles in the boundary layer, there have been virtually no studies that have examined this aspect of insect olfactory behavior, at least partly because it is difficult to study. However, in one relevant study of the feeding behavior of the caterpillar of *Manduca sexta* it was shown that removal of the antennae could alter the insect's response to cow pea, *Vigna sinensis*, from acceptance to rejection. Since the antennae bear olfactory organs it is almost certain that this insect is making its decision about this plant on the basis of the leaf odor. We believe that further investigations will reveal that close range odor cues are of widespread importance (see also Chapter 2).

4.2.3 The Plant Surface

The plant surface is covered by a layer of wax (see Fig. 2.1). The chemistry of the wax differs from plant to plant but the most commonly occurring components are alkanes, primary and secondary alcohols, and ketones (see Section 2.2). Sometimes, small amounts of sugars and amino acids leach out from within the leaf into the wax. These components are generally common to many different plants. In addition, the wax may contain chemicals that are characteristic of the plant. Volatile chemicals, either produced by the plant itself or adsorbed from the surrounding air, may also be present.

Many insects are known to respond to the chemical composition of the wax, or some of the chemicals occurring in it, and it is probable that this is a widespread phenomenon amongst phytophagous insect species. Wax chemicals may affect the establishment of insects on a plant, feeding behavior or oviposition behavior. The chemicals may have positive stimulatory effects, or they may be deterrent. As with odors and plant secondary compounds within the leaf, leaf surface chemicals cannot be categorized in any general sense with respect to their effects on insects. A compound that is stimulatory to one species may be deterrent to another.

It is sometimes suggested that the insects are really responding to chemicals within the leaf rather than those on the surface. Possibly these internal constituents are revealed by the insect abrading the leaf surface. However, specific searches have failed to reveal any signs of abrasion after insects have examined leaves, and experiments in which lipid extracts from plant surfaces, or specific chemicals, are applied to artificial substrates leave no doubt that phytophagous insects often do respond to chemical components of the leaf surface. The chemicals are generally perceived by the insect with its contact chemoreceptors on the tarsi or on the mouthparts. Female insects often have small numbers of contact chemoreceptors on the ovipositor and it might be supposed that these are also important in the detection of wax chemicals, although so far this has not been demonstrated. In a few cases wax chemicals may be volatile and so perceived by olfactory receptors. In the case of the swallowtail butterfly, *Papilio polyxenes*, electrophysiological recordings from tarsal sensilla have revealed the presence of neurons responding to the behaviorally significant chemicals.

There follow some examples of insect behavior that illustrate the roles of leaf-surface chemicals in host-plant selection.

Larvae of the diamondback moth, *Plutella xylostella*, feed on cabbage and other plants in the Brassicaceae. Some varieties of cabbage are, however, resistant to the insects. On resistant plants, newly hatched larvae spend more time walking and less time palpating than they do on a susceptible cultivar (Fig. 4.14). Eventually, the larvae leave the plant without feeding. This difference in behavior is due to differences in the surface waxes of the cultivars. The susceptible cultivar has the waxy bloom found on most cabbage plants; the resistant cultivar, however, has glossy leaves. When the waxes are removed from the leaf surface and replaced on glass, insects tested on these surfaces exhibit the same differences in behavior. The chemistry of the waxes is quite different. In particular, the resistant variety has a much greater proportion of primary alcohols than the susceptible one, and triterpenols are completely absent in the latter (Fig. 4.14c). The precise components that give rise to the behavioral differences have not yet been determined.

Leaf surface wax has also been shown to be important in the establishment of newly hatched larvae of the stem borer, *Chilo partellus*, on *Sorghum* and the potato tuber moth, *Phthorimaea operculella*, on potatoes.

Wax chemistry is also known to affect the food selection behavior of aphids, planthoppers, grasshoppers and caterpillars. When a grasshopper is about to feed it drums on the leaf surface with its maxillary and labial palps, an activity referred to as palpation. The tips of the palps carry large numbers, usually more than 100, of contact chemoreceptors, as well as a few olfactory receptors. When the insect palpates, it presumably gathers information about the surface chemistry of the plant and perhaps also the odors occurring close to the surface. The insects usually move as they palpate; this allows them to sample a larger area of the leaf than would be possible if they remained stationary. *Locusta migratoria* rejects many plants that are not acceptable as food at this point, though sometimes it will go on to bite the leaf. If the waxes are removed from these plants by briefly dipping the leaves in a lipid solvent the insect then bites the plant before rejecting it (Fig. 4.15).

One example that has been studied in greater detail concerns the response of *L.migratoria* to young *Sorghum*. Older leaves of this plant are completely

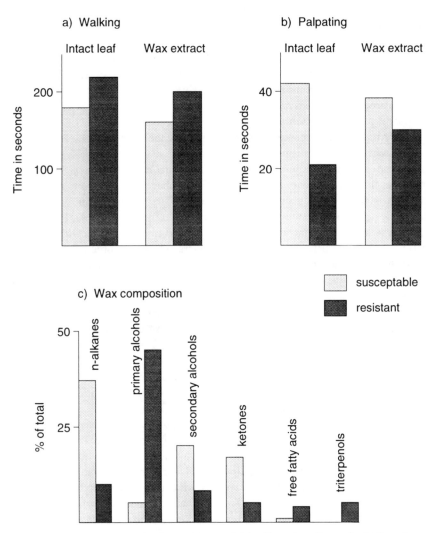

Figure 4.14. Response of neonate larvae of *Plutella xylostella* to the leaf surface waxes of susceptible and resistant cabbage cultivars. Figure shows responses on intact leaf and on a film of wax on a glass slide. Wax was extracted with hexane. **a)** Time walking. **b)** Time palpating. **c)** Composition of extracts from the two cultivars (after Eigenbrode et al., 1991).

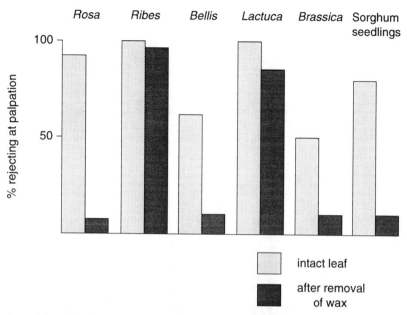

Figure 4.15. Effects of surface waxes on rejection of nonhosts by *Locusta migratoria*. Left-hand column shows the percentage of insects that rejected intact leaves without biting the leaf, i.e., after palpation alone. Right-hand column shows the response after leaf-surface wax was removed with a solvent. With *Ribes* and *Lactuca* there was no significant effect. This may have resulted from a failure to remove the wax or, especially in the case of *Ribes*, because rejection was mediated by odor (after Chapman, 1977; Woodhead, 1983).

acceptable, but the leaves of young plants of some cultivars are often rejected without feeding. The specific wax chemicals in a cultivar called 65D that are deterrent to *L.migratoria* are C_{19}, C_{21} and C_{23} alkanes, an ester of C_{12} acid with C_{24} alcohol, and *p*-hydroxybenzaldehyde. On young leaves of this cultivar, *p*-hydroxybenzaldehyde comprises 33% of the wax, esters 22%, free fatty acids 17%, and alkanes 13%. Not all the compounds in any one chemical class are deterrent. Fig. 4.16 shows that longer chain alkanes, which are relatively common in the wax of seedling plants, were not deterrent to *L.migratoria*.

In some cases wax chemicals are known to stimulate feeding. For example, the alcohols, hexacosanol (C_{26}) and octacosanol (C_{28}) are phagostimulatory for newly hatched silkworms, *Bombyx mori*.

Since ovipositing insects are unable to sample the inner contents of leaves it may be expected that they would make particular use of features of the leaf surface. Two species in which oviposition behavior has been extensively studied are the swallowtail butterfly, *Papilio polyxenes*, and the carrot fly, *Psila rosae*. Both species lay their eggs on the leaves of carrots as well as other species of

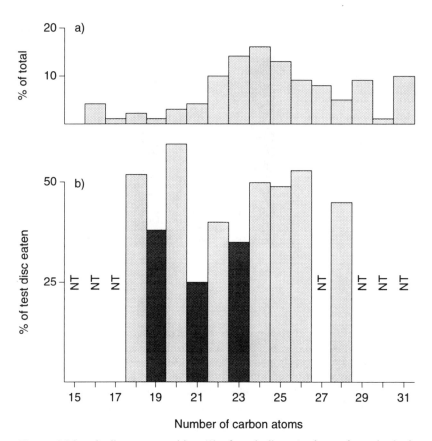

Figure 4.16. **a)** alkane composition (% of total alkanes) of wax from the leaf-surface of seedlings of *Sorghum* cultivar 65D. **b)** effects of different alkanes on feeding behavior of *Locusta migratoria*. Results of a choice test in which amounts eaten of glass-fiber discs with added sucrose are compared with discs with added sucrose plus an alkane. Fifty percent eaten indicates no preference for either control or test discs. Significantly less than 50% eaten (dark shading) indicates that the test disc was deterrent. NT = not tested (after Woodhead, 1983).

Apiaceae. The swallowtail uses several compounds to identify a suitable host plant. At least two of these are known to occur in the wax: chlorogenic acid, a compound found in most or all plants, and a glucoside of a luteolin flavonoid (see Section 2.4.2.3) which is specific to the Apiaceae. These compounds do not elicit oviposition when they are present singly; both must be present for egg-laying to occur. The compounds stimulate separate cells in the tarsal sensilla so that the information must be processed within the central nervous system before the insect is able to make a decision about the acceptability of a leaf. In addition to these two compounds, three others are also necessary to produce oviposition at a level equivalent to that occurring on the host plant.

In the carrot fly, the only single compound that induced oviposition at concentrations found on the carrot leaf was a polyacetylene called falcarindiol. This compound only occurs in Apiaceae. However, when it was presented alone in experiments it was less effective than a fraction from the plant surface that contained a mixture of compounds. It was found that five other compounds, although present on the leaf surface at concentrations below that at which they stimulated oviposition when presented singly, nevertheless contributed to the effectiveness of falcarindiol when presented with it in a mixture.

These two examples of oviposition illustrate two points. First, although the insects oviposit on the same plant, different chemicals provide the cues for identification. Secondly, in both cases a mixture of compounds is important; single components are either ineffective or less effective than a mixture. This is not always true, however. The oviposition behavior of several *Brassica*-feeding insects has been studied and, in these, glucosinolates on the leaf surface play a key role. In two species of *Pieris* one compound, glucobrassicin, is much more effective than any other and, when applied to an artificial substrate, produces as much oviposition as do *Brassica* plants. Glucosinolates are also oviposition stimulants for the diamondback moth, *Plutella xylostella*, another crucifer-feeding specialist, but in this case no one chemical is particularly important; the response is similar to any one of a range of different glucosinolates. Glucosinolates are water soluble. The experiments with glucobrassicin suggest that this compound, at least, is bound in the leaf surface and so would not be readily removed by rain.

In the examples given so far of the effects of leaf surface chemicals on behavior it is likely that the responses are innate, although this has not always been fully established. But there are also examples in which insects modify their responses to surface chemicals as a result of experience.

When a nymph of *Locusta migratoria* first encounters a nonhost plant it will often bite it before rejecting it. Presumably rejection is based on the presence of deterrent compounds within the leaf that are released when the insect bites. On subsequent encounters, however, the insect will often reject the plant without biting, when it has only palpated on the leaf surface. By the fourth or fifth encounter, most insects reject without biting (Fig. 4.17). Clearly they are responding to properties of the leaf surface that, initially, were not important to them. This has been interpreted as associative learning, the insect associating the internal deterrent properties of the leaf with its external properties. The possibility that this behavior could also arise from sensitization (Section 6.3) has not been excluded. Similar observations have been made of *Schistocerca americana*. In neither of the two species have the specific properties of the leaf surface that are important been examined.

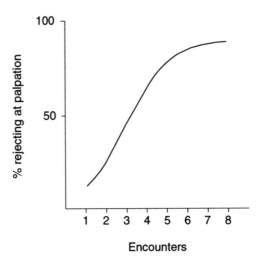

Figure 4.17. Responses of *Locusta migratoria* on successive encounters with the nonhost, *Senecio vulgaris*. Initially, most insects bit the leaf before rejecting it. With each successive encounter, a greater percentage rejected following palpation without biting (after Blaney and Simmonds, 1985).

4.2.4 The internal constituents of the leaf

When an insect bites into a leaf it releases the cell contents and, we believe, these flow over the mouthparts and stimulate the contact chemoreceptors. In grasshoppers, tracts of noninnervated hairs on the innerside of the mouthparts appear to direct the plant fluids to the receptors (Fig. 4.18). These hairs are easily wetted by water while the rest of the cuticle resists wetting. Functionally similar adaptations probably exist in other insects although they have not been investigated.

The plant cell contains large numbers of different chemicals, many of which have the capacity to stimulate the contact chemoreceptors on the mouthparts. Some will be phagostimulatory and others deterrent. It is convenient and necessary to consider them separately, but we emphasize again that the insect response depends on the overall sensory input and may not reflect the sum of the responses to the individual components.

4.2.4.1 Phagostimulants

The principal phagostimulants are nutrients, and especially sugars. In general, the same sugars are stimulating for different species (Table 4.5), sucrose and fructose generally being the most effective. Pentose sugars are not usually stimu-

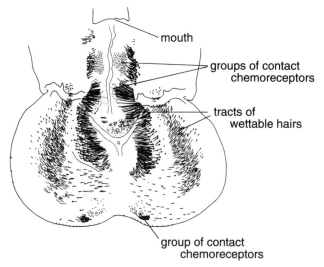

Figure 4.18. Inside of the labrum of a grasshopper, showing the positions of tracts of wettable hairs. It is probable that these tracts conduct plant sap to the groups of contact chemoreceptors just outside the mouth (after Chapman and Thomas, 1978).

lating. In all cases, the effectiveness of the sugar, as measured by the meal size or the amount of an artificial substrate consumed over a period, increases with its concentration (Fig. 4.19a) at least within the limits occurring in most plants.

Despite the importance of proteins nutritionally, there is, as yet, no good evidence that insects can taste protein. They can, however, usually taste some amino acids (Table 4.5), although the stimulating power of these compounds is usually low compared with sugars (Fig. 4.19a). Consequently, for most insects feeding on most plants, phagostimulatory effects are likely to be dominated by sugars. In a caterpillar and a grasshopper, it has been shown that a sucrose concentration equal to that occurring in host plants is sufficient to make the insect eat a maximum-sized meal on an otherwise neutral substrate. In addition, it has been shown that the amount of feeding on leaves is correlated with their sugar content (Fig. 4.20). Such examples do not prove that sugar is the only factor affecting feeding, but they indicate that the sugar content of leaves is likely to be very important in food selection. However, amino acids may affect meal duration or the intervals between meals, and insects are able to distinguish artificial diets that are high in protein from those that are not by associative learning (Section 6.4).

Phospholipids and some nucleotides may also be phagostimulatory (Table 4.5), although the latter are much less important for phytophagous insects than for blood-sucking insects. Inorganic salts, which are essential nutrients, usually

Table 4.5. *Phagostimulatory effect of some nutrient compounds tested singly for some phytophagous insects.* + = *weakly stimulating,* +++++ = *strongly stimulating,* — = *no effects,* • = *not tested.*

Compounds	Locusta migratoria (Orthoptera)	Aphis fabae (Hemiptera)	Oncopeltus fasciatus (Hemiptera)	Leptinotarsa decemlineata (Coleoptera)	Pieris brassicae (Lepidoptera)
Carbohydrates					
Pentoses					
D-ribose	—	•	•	—	—
D-xylose	—	•	•	—	—
Hexoses					
D-fructose	+++++	•	•	+	—
D-glucose	+++	+	•	+	++
L-sorbose	+	+	•	—	—
Disaccharides					
D-cellobiose	—	•	•	—	—
D-maltose	+++++	+	•	—	—
D-sucrose	+++++	+++++	•	+++++	+++++
Sugar Alcohols					
Inositol	+	•	•	—	—
Sorbitol	+	—	•	—	—
Amino Acids					
L-alanine	•	++	++	++	•
γ-aminobutyric acid	—	•	+	++	—
L-methionine	+	++	++	—	—
L-serine	+	++	+	++	—
L-phenylalanine	—	—	+	—	—
L-tyrosine	—	•	•	—	•
L-proline	+	++	•	++	+
Nucleotides					
AMP	—	•	•	—	•
ADP	—	•	•	—	•
ATP	—	•	•		•

130

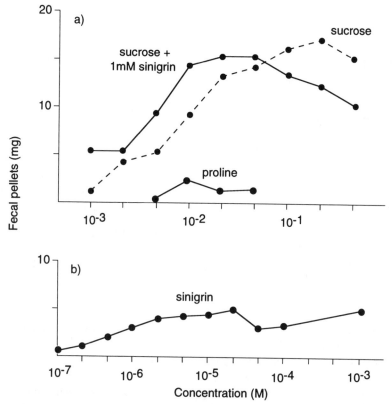

Figure 4.19. Responses of caterpillars of *Pieris brassicae* to some phagostimulants. The amount eaten was determined by measuring the dry weight of fecal
pellets produced in a 24-hour period. Insects were fed on blocks of agar/cellulose
to which the test chemical was added (after Blom, 1978).

have no effect at the concentrations at which they occur in plants. At higher
concentrations they become deterrent.

These various compounds will be present in all plants and will contribute to
a plant's acceptability. But they cannot provide the information necessary to
confer specificity and no examples are known of host-plant specificity based on
the presence of a particular nutrient or a particular combination of nutrients.

There are also many examples of plant secondary compounds acting as phagostimulants. In some cases these are widely-occurring compounds that affect
insects feeding on a range of different plants. The flavonoid glycoside rutin is
an example. It occurs in many different plant families and stimulates feeding in
polyphagous species, like the larva of *Helicoverpa zea* and the grasshopper,
Schistocerca americana. Other examples are listed in Table 4.6. It is often true

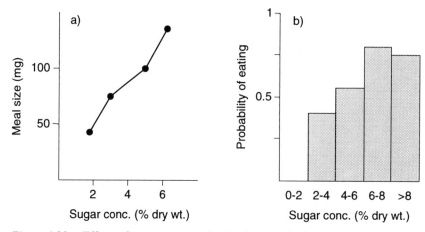

Figure 4.20. Effects of sugar concentration in plants on feeding. **a)** Meal sizes taken by fifth instar nymphs of *Schistocerca gregaria* on seedling wheat (EAB, unpublished). **b)** The probability that larvae of *Tyria jacobaeae*, the cinnabar moth, would feed on leaves with different sugar levels. There was no difference in the probability of feeding in relation to protein levels in the same leaves (after Soldaat, 1991.)

that these phagostimulatory effects are only observed when the compounds are present in low concentrations.

In a many other cases, however, particular secondary compounds are only found in one or a small number of plant taxa. In these cases, the chemicals can provide indicators or sign stimuli to a monophagous or oligophagous insect that it is on the correct host and so help to define its host range. For example, the caterpillars of the butterfly subfamily Pierinae feed almost exclusively on cruciferous plants. These are characterized by glucosinolates which have been found to be phagostimulatory for larvae and oviposition stimulants for females of Pierinae as well as for many other insects that are oligophagous on these plants. Iridoid glycosides, which are monoterpenoids, characterize the host plants of the buckeye butterfly, *Junonia coenia,* and the checkerspot butterflies, *Euphydryas* spp., and are phagostimulants and oviposition stimulants for them. There are many other examples of insect genera or species in which phagostimulatory effects are produced by chemicals that are characteristic of the host plant (Table 4.7).

Sometimes the chemicals are phagostimulatory by themselves. This is true of some glucosinolates for crucifer-feeding insects and of populin for the beetle, *Chrysomela vigintipunctata.* In other cases, the chemical may have no effect by itself, but may synergise feeding on an artificial diet containing sugar. Fig. 4.21 shows an example of this for sinigrin and the diamondback moth, *Plutella xylostella.* It is also often true that the addition of an appropriate sign stimulus chemical to a leaf of an unacceptable plant will cause an insect tuned to this substance to eat the treated leaf.

Table 4.6. Some plant secondary compounds that are produced by plants in many different families and which are phagostimulatory for some insects. In some cases phagostimulation only occurs when the chemical is present in low concentrations.

Chemical	Chemical class	Insects that are phagostimulated
Anthraquinone	quinone	*Schistocerca gregaria* (grasshopper)
Caffeic acid	phenolic acid	*Bombyx mori* (caterpillar)
Chlorogenic acid	phenylpropanoid acid	*Leptinotarsa decemlineata* (beetle); *Bombyx mori* (caterpillar)
Chrysophanol	quinone	*Schistocerca gregaria* (grasshopper)
Cinnamic acid	phenylpropanoid	*Schistocerca gregaria* (grasshopper)
Quercitrin	flavonoid glycoside	*Anthonomus grandis* (beetle); *Bombyx mori* (caterpillar)
Linamarin	cyanogenic glycoside	*Epilachna varivestis* (beetle)
Luteolin-7-glucoside	flavonoid glycoside	*Chrysomela vigintipunctata* (beetle)
Rutin	flavonoid glycoside	*Schistocerca americana* (grasshopper); *Plagiodera versicolora* (beetle); *Heliothis virescens* (caterpillar)
Tannic acid	tannin	*Anacridium melanorhodon* (grasshopper); *Lymantria dispar* (caterpillar)

Table 4.7. Some plant secondary compounds that are taxon specific and phagostimulatory for some insects. It is believed that these compounds have a major role in defining the host ranges of the insects named.

Chemical	Chemical class	Plant taxon	Insects that are phagostimulated
Catalpol	monoterpenoid	*Catalpa*	*Ceratomia catalpae* (caterpillar)
Catalpol	monoterpenoid	plantains	*Euphydryas chalcedona* (caterpillar)
Cytisine	alkaloid	broom	*Uresiphita reversalis* (caterpillar)
Gossypol	sesquiterpene	cotton	*Anthonomus grandis* (beetle)
Hypericin	quinone	St. John's wort	*Chrysolina brunsvicensis* (beetle)
Monocrotaline	alkaloid	*Crotalaria*	*Utetheisa ornatrix* (caterpillar)
Morin	flavonoid	mulberry	*Bombyx mori* (caterpillar)
Nordihydroguaiaretic acid	phenylpropanoid	creosote bush	*Bootettix argentatus* (grasshopper)
Phloridzin	flavonoid	apple	*Aphis pomi* (aphid)
Populin	phenolic	willow	*Chrysomela vigintipunctata* (beetle)
Salicin	phenolic	willow	*Plagiodera versicolora* (beetle); *Laothoe populi* (caterpillar)
Sinigrin	glucosinolate	cabbage family	*Brevicoryne brassicae* (aphid); *Phyllotreta* (beetle); *Pieris brassicae* (caterpillar); *Athalia proxima* (sawfly larva)
Sparteine	alkaloid	broom	*Acyrthosiphon spartii* (aphid)

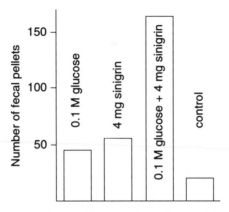

Figure 4.21. The synergistic effect of sinigrin with glucose. Larvae of *Plutella xylostella* were fed on cellulose-agar gels to which glucose, sinigrin or both were added. Glucose or sinigrin alone had slight phagostimulatory effects compared with the control (cellulose-agar alone). A mixture of the two in the same concentrations greatly enhanced the amount eaten over an 18-hour period. Amount eaten is expressed as the number of fecal pellets produced by 10 larvae (after Nayar and Thorsteinson, 1963).

These data leave no doubt that such chemicals have a special significance to the insects and play important roles in host-plant recognition. Probably, in a majority of instances, they provide a principal basis for monophagy and oligophagy. However, precisely how they function in host-plant selection is not clear. For most phytophagous species examined, sugars are major phagostimulants and the insects will often eat sugar-based diets even in the absence of the host-identifying chemical. For example, in *Pieris brassicae* glucosinolates clearly have a phagostimulatory effect. However, the effect of sinigrin on food intake over a 24 hour period is small compared with the effects of sucrose alone (Fig. 4.19). Sucrose at concentrations above 5×10^{-2} M, approximately that occurring in the plant, maximizes food intake; the insect cannot eat any more so that the addition of sinigrin can have no effect. It only has an effect on food intake when the concentration of sucrose is low. When the effects of sinigrin on first meal length and first feeding bout are considered, instead of food intake over 24 hours, it is found to have no effect when mixed with 3×10^{-1} M sucrose, but some effect was observed with 3×10^{-3} M sucrose. The sugar level is probably never as low as this in the plant. Most insect/plant relations have not been investigated with the thoroughness of the *Pieris/Brassica* association, and there is no case in which we have a clear idea of how the supposed sign stimulus has its effect.

However, it may be true that in whole plants the effects of host-specific chemicals are synergised by other constituents that enhance their phagostimulatory effect well above that of the nutrient phagostimulants. This thesis has been argued in particular in the case of a number of beetles feeding on Brassicaceae. For example, sinigrin alone is not a phagostimulant for the weevil, *Ceutorhynchus constrictus*, but it is highly effective when presented together with other chemicals from the host plant which are believed to be flavonoids. Perhaps the combined phagostimulatory effect is greater than that of the nutrients in the plant, although this has not been investigated. The possible importance of combinations of chemicals is also suggested by the increasing number of studies on oviposition behavior that demonstrate synergistic effects between a number of different compounds (Section 4.2.3). Comparably detailed studies on phagostimulants have not generally been undertaken.

These effects could arise through the interaction of chemicals at the insect's sensory receptors (see Section 3.2.3) so that the sign stimulus dominates the information that the insect receives. So far only one study in which the activities of different neurons have been distinguished has investigated the neurophysiological responses of oligophagous insects to a range of saps from host and nonhost plants. One of the species examined in this study was the Colorado potato beetle, *Leptinotarsa decemlineata*, which is oligophagous on a number of solanaceous plants. No particular group of chemicals is known to provide sign stimuli for this species, but the results of the study are important in the current context.

Four species of Solanaceae were examined: *Solanum tuberosum* (potato) and *Solanum dulcamara* (nightshade), which were readily accepted as food, and *Solanum elaeagnifolium* (horse nettle) and *Lycopersicon esculentum* (tomato), which were less acceptable. Stimulation with the sap of *S. tuberosum* usually produced a response in several different cells. One of these (cell 1 in Fig. 4.22a) fired whenever the stimulus was applied; the other cells were much less consistent. In addition, the firing rate of cell 1 was much higher than that of any of the other cells (Fig. 4.22c). A similar pattern was observed when *S. dulcamara* was the stimulus, but with the other two plants the pattern was different. With *L.esculentum*, none of the cells fired consistently (Fig. 4.22b) and none fired at a high rate. With *S. elaeagnifolium*, the pattern was different again. Cell 1 fired consistently at a high rate, much as it did with the more acceptable plants, but cell 2 was also active and fired at a relatively high rate.

The pattern of activity that distinguishes highly acceptable from less acceptable plants is the high and consistent firing rate of cell 1 in the absence of significant activity of other cells. It has been suggested that, in addition, the high variability of the responses of the other cells is itself used as a signal. The insect regards a high level of variability in a sequence of stimulations as "foreign," or not acceptable.

This type of interpretation did not fit as well to the behavior of two other species of *Leptinotarsa* examined, but these results are based on an investigation of only one sensillum. If other sensilla on the mouthparts had been examined,

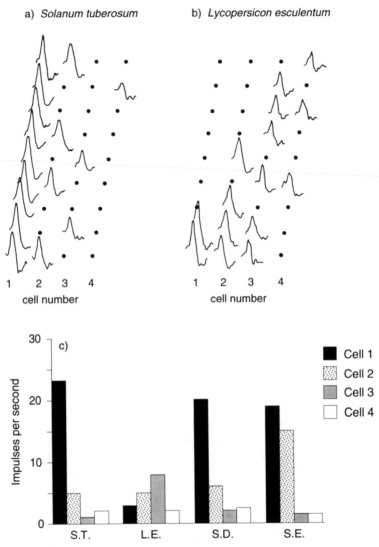

a) *Solanum tuberosum*   b) *Lycopersicon esculentum*

Figure 4.22. Responses of sensory neurons in the galeal sensillum of adult *Leptinotarsa* to the sap of **a)** potato, which is eaten readily, and **b)** tomato, which is less readily eaten. The different neurons within the sensillum are categorized by the shapes of the action potentials. Each horizontal line represents a single stimulation (there were nine stimulations for each plant sap). Cell one always fired when the sensillum was stimulated with potato sap. No other cell responded regularly to either of the saps. **c)** The average number of action potentials produced in the first second of stimulation by saps of acceptable (*Solanum tuberosum* (S.T.) and *S. dulcamara* (S.D.)) and less acceptable (*Lycopersicon esculentum* (L.E.) and *S. elaeagnifolium* (S.E.)) plants (after Sperling and Mitchell, 1991).

a better correlation between sensory physiology and behavior might have been obtained.

4.2.4.2 Pharmacophagy

There are some cases in which a secondary chemical is highly phagostimulatory to an insect in the complete absence of other chemicals and where it has some function independent of nutrition. Insects exhibiting this phenomenon are said to be pharmacophagous. The insects in which this habit has been demonstrated so far usually search for this particular chemical independently of their main food supply.

Amongst phytophagous insects, pharmacophagy has been clearly demonstrated in only a few instances, although it may prove to be much more widespread. Some examples relate to the sequestration of chemicals by the insects for defense against predators; in other cases, the chemicals are used for pheromone production. In some cases, the chemicals used are obtained from plants outside the principal host range of the insect; in other cases they occur in commonly used hosts. As we become familiar with more examples of insects that make specific uses of secondary compounds, we may find that pharmacophagy represents one extreme of a spectrum which grades into those situations where the specific chemical is also a normal component of the food.

Larvae of the turnip sawfly, *Athalia rosae*, feed exclusively on cruciferous plants. Adults, however, visit an entirely unrelated plant, *Clerodendrum trichotomum*, in the plant family Verbenaceae, and feed on the leaf surface. This plant contains a group of terpenes called clerodendrins and one of these, clerodendrin D, is a phagostimulant for the adult sawflies.

Adults of the western spotted cucumber beetle, *Diabrotica undecimpunctata*, are polyphagous, although they are often found on plants in the family Cucurbitaceae. Cucurbitacins from these plants are highly phagostimulatory for this species as well as many other diabroticine beetles, although the insects do feed on many plants that do not contain cucurbitacins. The addition of a cucurbitacin to a leaf of soybean, which is not normally eaten by *Diabrotica*, causes the beetles to feed actively, and *D.undecimpunctata* can detect as little as 3 ng of cucurbitacin pipetted on to a chromatography plate.

A similar situation occurs in grasshoppers of the genus *Zonocerus*. These insects are highly polyphagous. However, pyrrolizidine alkaloids are powerful phagostimulants for them and the insects will feed on filter papers containing no other added chemicals (Fig. 4.23). At times they also feed actively on flowers containing the alkaloids. The insects sequester the chemicals.

Clerodendrins, cucurbitacins and pyrrolizidine alkaloids are sequestered by the insects feeding on them. This is despite the fact, in the case of *A.rosae*, that clerodendrin D is only a minor component of the *Clerodendrum* leaf. All the compounds are known to be deterrent to other insects and to be distasteful to

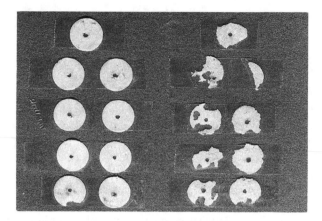

Figure 4.23. Glass fiber discs impregnated with a pyrrolizine alkaloid (right) and with solvent only (left) after being exposed to *Zonocerus elegans* in the field (after Boppré et al., 1984).

vertebrates, and this is probably their role in these insects, although this has not been proven.

The use of secondary chemicals as precursors of pheromone occurs in the Lepidoptera. For example, the adults of some danaid butterflies seek out plants containing pyrrolizidine alkaloids and can often be seen extracting the chemicals from the surfaces of dead leaves or exposed roots. They do this by regurgitating fluid from the proboscis to dissolve the chemicals and then sucking the fluid up again. These alkaloids are used as precursors of aphrodisiac pheromones.

The most extreme instance of pharmacophagy known so far concerns moths in the genus *Creatonotus*. These insects also require pyrrolizidine alkaloids to produce a male pheromone, but, unlike the danaids, the alkaloids are obtained with the larval food. However, in this case, the alkaloids also stimulate the production of a structure called the corema (plural coremata) at the end of the male abdomen. This structure is everted by the male during courtship and it appears to be the site of production and release of the male pheromone. Insects that feed as larvae on food that lacks pyrrolizidine alkaloids not only do not produce the male pheromone but also have greatly reduced coremata when they become adult. This is the only instance known so far where pharmacophagy is associated with morphogenesis.

4.2.4.3 Deterrents

Although there are many instances of plant secondary compounds acting as phagostimulants, their more usual role is to inhibit or deter feeding and they are commonly called feeding deterrents. There are many examples of this type of action. Their effectiveness in inhibiting feeding increases as their concentration is increased relative to amount of phagostimulant present (Fig. 4.24).

Most plants contain deterrents for most insects. Table 4.8 demonstrates this for a number of different beetles and species of Lepidoptera, ranging from monophagous to polyphagous. Irrespective of the feeding pattern of the insect, all plants that were not acceptable contained deterrents. Fig. 4.25 shows the deterrent effects of a range of plants tested against *Locusta migratoria*, an oligophagous species feeding mainly on grasses, and *Schistocerca gregaria*, a polyphagous species. In *L. migratoria*, a full meal was about 100 mg. Plants that were not eaten extensively were usually rejected without any feeding; very few plants were eaten in intermediate amounts. All the plants that were rejected yielded at least one extract that was deterrent and sometimes it was evident that different classes of deterrent compounds were present even within one plant. Plants that were eaten in large quantities usually did not contain a deterrent for this insect, but this was not always true.

With *S. gregaria,* the picture is different although many of the same plants were tested. Some plants were rejected without feeding and some were eaten in large amounts, up to 200 mg in a meal, but many plants were eaten in intermediate amounts. All the plants that were rejected or eaten in small amounts contained deterrents, but most of the others did not.

These examples illustrate several general conclusions that can be drawn from the many studies on insect deterrents:

1. Deterrent effects are very widespread and all nonhosts contain substances which act as deterrents

2. Plants differ in their deterrence towards different insect species and oligophagous insects are deterred by more plants than polyphagous species

3. Even plants that are acceptable may contain deterrents.

Arising from the first of these conclusions is the probability that many different types of chemical compounds are involved in deterrence. Some of these are listed in Table 4.9. The examples chosen are limited by the fact that few studies have examined the effects of a wide range of compounds on one species of insect. In the case of *L.migratoria*, over 100 compounds are known to have deterrent effects on feeding. Probably this species is deterred by hundreds or even thousands of different compounds. This is probably true for most phytophagous insects, although we do not yet have the data to prove this point. It is true, however, that a very large number of plant secondary compounds has been shown to be deterrent to a variety of insects. Much of this knowledge has been obtained in the search for compounds in plants that might be used in plant protection. Although these data are not directly related to host-plant selection, they demon-

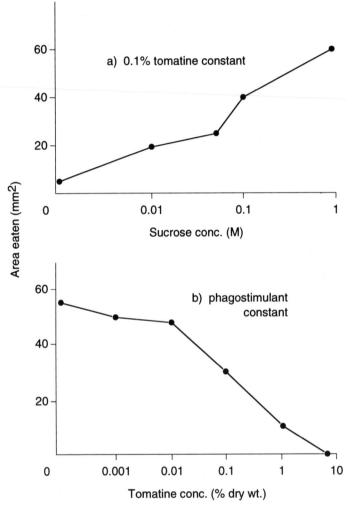

Figure 4.24. Interactions of phagostimulant and deterrent chemicals, showing the amounts of wheat-flour wafers eaten by *Locusta migratoria*. **a)** Keeping the deterrent (tomatine) constant, while increasing the concentration of sucrose. **b)** Increasing the concentration of deterrent, while keeping the level of phagostimulant constant (after Bernays and Chapman, 1978).

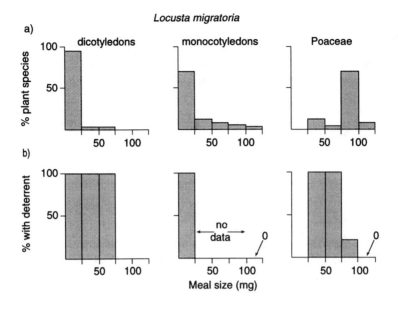

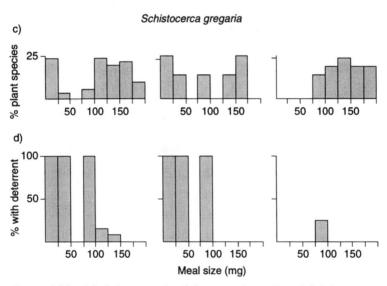

Figure 4.25. Meal sizes eaten by **a)** *Locusta migratoria* and **c)** *Schistocerca gregaria* on a range of plants. **b)** and **d)** show the percentage of plants in each category that yielded an extract that was deterrent to the insects when presented to them on a stimulating, wheat-flour substrate. Extracts were made with three different solvents expected to be solvents for different chemicals (partly after Bernays and Chapman, 1978).

Table 4.8.   Widespread occurrence of deterrent effects in nonhost plants for a number of different insect species. In this study leaf discs from plants belonging to a wide range of families were presented to the insects sandwiched between discs of the normal host plant. The amount of this test sandwich that was eaten was compared with the amount of a control sandwich eaten. In the control sandwich, all three discs were from the normal host plant. This method insures that a failure to feed is due to a deterrent effect and does not simply result from the lack of a specific phagostimulant (after Jermy, 1966).

Species	Diet breadth	Numbers of plants tested	% Eaten	% Plants with deterrents
Coleoptera (adults)				
Tanymecus dilaticollis	polyphagous	21	57	43
Phyllobius oblongus	polyphagous	41	27	73
Leptinotarsa decemlineata	oligophagous	83	8	92
Cassida nebulosa	monophagous	30	10	90
Phytodecta fornicata	monophagous	60	5	95
Lepidoptera (larvae)				
Hyphantria cunea	polyphagous	56	45	55
Pieris brassicae	oligophagous	75	16	84
Plutella xylostella	oligophagous	62	6	94
Hymenoptera (larvae)				
Athalia rosae	oligophagous	40	23	77

strate the variety and widespread occurrence of deterrents (see Morgan and Mandava, 1985, for an extensive listing of deterrent compounds).

The fact that oligophagous species are deterred by more plants than polyphagous species implies that they are deterred by more chemicals. This proves to be true, not because there is a qualitative difference in the types of chemicals that affect the different insects, but because oligophagous insects are more sensitive to deterrents. This is illustrated in Fig. 4.26 for two chemicals and several grasshoppers with different host-plant ranges. The figure also shows that a chemical may be phagostimulatory for one species and deterrent for others.

The difference between species with different host-plant ranges is further illustrated in Fig. 4.27 where two pairs of species, one oligophagous on grasses, the other polyphagous, are compared. The grass-feeding insect is nearly always deterred by a lower concentration of any chemical than the polyphagous one. The polyphagous species is affected by all the same chemicals, but sometimes the concentration needed to produce an effect is well beyond that likely to occur in nature.

It is generally recognized that many compounds are deterrent to "nonadapted" species, that is species that do not normally feed on a plant or polyphagous species for which a particular plant is only one of many hosts. However, many plants that are eaten contain compounds that are deterrent even for the species that feed on them regularly. This is illustrated in Fig. 4.26b. *Ligurotettix coquilletti*

Table 4.9. Some plant secondary compounds that are deterrent to insects from three different orders. Compounds are listed in the order that they appear in Chapter 2. Gaps in the table do not imply that these compounds are not deterrent; they result from lack of data.

Chemical Class	*Locusta migratoria* Orthoptera Oligophagous	*Dysdercus fulvoniger* Hemiptera Polyphagous	*Spodoptera*[1] Lepidoptera Polyphagous
non-protein amino acid	L-canavanine		
amine	hordenine	hordenine	
alkaloid	nicotine	nicotine	isoboldine
cyanogenic glycoside	dhurrin	amygdalin	
phenol	*p*-hydroxybenzoic acid	salicin	
phenylpropanoid	coumarin		angelicin
flavonoid	vismione	cyanidin	vismione
tannin	tannic acid	tannic acid	
monoterpene	carvone		carvone
sesquiterpene	gossypol		absinthin
diterpene	abietic acid		clerodendrin
triterpene	azadirachtin	betulin	azadirachtin
glucosinolate	sinigrin		

[1]Data for *Spodoptera* are for *S. littoralis* and *S. litura,* both of which are polyphagous.

habitually feeds on *Larrea*, yet nordihydroguaiaretic acid (NDGA) is deterrent to it at the concentrations normally found in the plant. The insect normally eats older leaves with a lower content of NDGA and males tend to select bushes with relatively low concentrations of NDGA as the foci of their territories. Nevertheless, the concentration of NDGA in foliage that they consume is still probably within the range which, experimentally, is deterrent for this species.

Less restricted species, like the Colorado potato beetle, also eat plants containing deterrents. For example, *Solanum dulcamara* is readily eaten despite the fact that its characteristic alkaloid, soladucine, is deterrent to the beetle. The same is true of tomato, *Lycopersicon esculentum*, although in this case the characteristic alkaloid, tomatine, is highly deterrent to the beetle and the plant is less readily eaten (see Table 4.10).

Similar data exist with respect to oviposition. The monarch butterfly, *Danaus plexippus*, lays its eggs on species of milkweed, *Asclepias*. These plants are characterized by containing cardiac glycosides (cardenolides) which are sequestered by the caterpillars and make the adult butterflies distasteful to vertebrate predators. These chemicals are also oviposition stimulants for the females and must be present in a plant to initiate oviposition. Nevertheless, fewer eggs are laid on plants with high concentrations of cardenolides which are deterrent to the ovipositing females. In this instance, the females apparently balance these two conflicting effects by preferring to oviposit on plants with intermediate amounts of cardenolide.

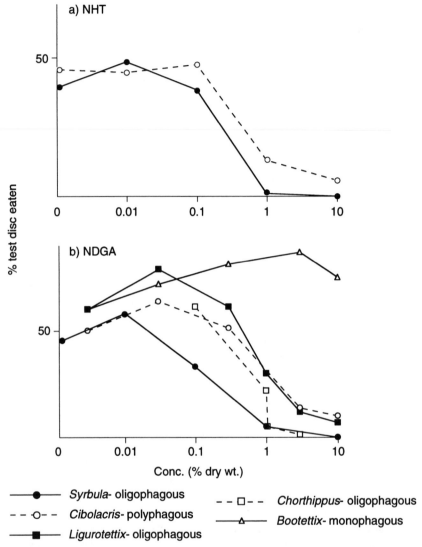

Figure 4.26. Effects of two plant secondary compounds, **a)** nicotine hydrogen tartrate (NHT), and **b)** nordihydroguaiaretic acid (NDGA) on the feeding behavior of grasshoppers. All are from the same subfamily, Gomphocerinae. *Syrbula montezuma* and *Chorthippus curtipennis* are both grassfeeders, *Cibolacris parviceps* is polyphagous, *Ligurotettix coquilletti* is oligophagous and commonly feeds on *Larrea*. *Bootettix argentatus* is monophagous on *Larrea*. NDGA is characteristic of *Larrea*. Values greater than 50% indicate phagostimulation; less than 50% indicates deterrence (partly after Chapman et al., 1988).

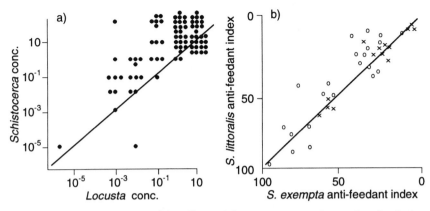

Figure 4.27. Comparisons of the effects of deterrent compounds on pairs of polyphagous and oligophagous (graminivorous) insects. **a)** *Schistocerca gregaria* (polyphagous) and *Locusta migratoria* (graminivorous), two locust species. Axes show the concentrations (dry weight on a wheat flour wafer) required to reduce feeding on the substrate by 50%. Each point represents a different compound (after Bernays and Chapman, 1978). **b)** *Spodoptera littoralis* (polyphagous) and *Spodoptera exempta* (graminivorous), two caterpillars. Axes show a deterrent index, based on choice tests with glass fiber discs treated with sucrose, with or without the test chemical. A value of 100 indicates complete deterrence; a value of zero (0) indicates no effect. Each "x" represents a different drimane, each "o" a different clerodane diterpenoid. In a) and b) the diagonal line shows the result to be expected if compounds were equally deterrent to both species. Points above the line show that a higher concentration of the compound is required to cause deterrence in the polyphagous species (after Blaney et al., 1987,1988).

These examples illustrate the point that feeding or ovipositing on plants that contain deterrents occurs in a variety of insect/host-plant associations. It is probably widespread and may even be a general phenomenon. We believe that this behavior reflects the balance that normally occurs between the deterrent and phagostimulatory components of plants. This balance will vary with different insects so that a plant species may be generally acceptable to one herbivore species and completely unacceptable to another. However, a shift in favor of the deterrent may make the plant unacceptable even to the species that normally feeds on it. This shift could occur as a result of an increase in the concentration of deterrent or a decrease in the quantities of phagostimulants.

It is commonly true, however, that "adapted" insects, monophagous or oligophagous species that are associated with a specific plant species or group of species, are relatively insensitive to the secondary compounds that characterize their host plants. A good example of this occurs in caterpillars of the small ermine moths of the genus *Yponomeuta*. Several European species are monophagous on different woody plants. In most species, deterrent cells in the galeal sensilla (see Fig. 3.4) are sensitive to the compound phloridzin although their normal hosts

Table 4.10. Responses of four species of beetles in the genus Leptinotarsa to different alkaloids and to the plants in which the alkaloids occur. "Response to alkaloid" shows the response to the alkaloids tested on an artificial substrate; "Response to plant" indicates whether or not the intact leaves are eaten (after Hsaio, 1974, 1988).

Alkaloid	Plant containing alkaloid	Response to alkaloid[1]				Response to plant[2]			
		decemlineata	haldemani	juncta	texana	decemlineata	haldemani	juncta	texana
Demissine	*Solanum jamesii*	——	——	——	——	—	—	—	—
l-hyoscyamine	*Datura*	0	——	——	——	—	—	—	—
Solanine	*Solanum tuberosum*	0	—	——	——	+++	+++	—	+
Tomatine	*Lycopersicon*	——	0	——	——	+	+++	—	—
Soladucine	*Solanum dulcamara*	——	No data	No data	No data	+++	No data	No data	No data

[1] 0 = no effect; — = slightly deterrent; —— = strongly deterrent.

[2] — = not eaten; + = some feeding; +++ = readily eaten.

do not contain this compound. However, *Y. malinellus* feeds on apple which does contain phloridzin, but its deterrent cells are insensitive to this compound at the concentrations tested (Table 4.11).

Similarly, amongst the beetles of the genus *Leptinotarsa*, it is common for species to be insensitive to the alkaloids occurring in their primary hosts, although the same compounds are deterrent to other members of the genus (Table 4.10).

Finally, it is often true that compounds that are deterrent for most insect species are stimulating for species that habitually feed on plants containing them. This is illustrated in Fig. 4.26. Nordihydroguaiaretic acid (NDGA) is phagostimulatory for *Bootettix argentatus* at all concentrations although it is strongly deterrent for all other species tested. Other examples of secondary compounds acting as phagostimulants or oviposition stimulants for adapted species are given in Table 4.7 (and see Section 4.2.3). All the compounds listed are known to be deterrent for many nonadapted species.

4.2.5 Making the unpalatable palatable

Plant secondary compounds are not uniformly distributed through leaf tissue (see Section 2.4) so that it is sometimes possible for an insect to avoid them by eating only those parts of a leaf from which the compounds are absent. This is illustrated by first instar larvae of *Heliothis virescens* feeding on cotton. Gossypol, which is produced in glands in the leaves of cotton, is deterrent to *Heliothis* larvae. Larger larvae cannot avoid damaging the glands as they feed, but first instar larvae can eat round the glands, apparently without damaging them. Each gland is surrounded by an envelope of cells containing anthocyanins (see Fig. 2.6a)

Table 4.11. *Sensitivity of larvae of small ermine moths,* Yponomeuta, *to phloridzin.* + *in the sensory response indicates that the deterrent cell in the medial sensillum of the galea is active (after van Drongelen, 1979).*

Species	Host Plant	Phloridzin in host[1]	Behavioral response to phloridzin[2]	Sensory response to phloridzin[3]
cagnagellus	*Euonymus*	—	—	+
evonymellus	*Prunus*	—	—	+
irrorellus	*Euonymus*	—	·	+
mahalebellus	*Prunus*	—	·	+
malinellus	*Malus*	+	·	—
padellus	*Prunus*	—	·	+
plumbellus	*Euonymus*	—	·	+
rorellus	*Salix*	—	·	+
vigintipunctatus	*Sedum*	—	·	+

[1] + = present; — = absent

[2] — = deterrence; · = no data

[3] + = response; — = no response

and it is possible that the insects are deterred by the anthocyanins so that they do not touch the glands.

In other cases, insects are known to make leaves palatable by reducing the production of deterrents. When an individual of the variegated grasshopper, *Zonocerus variegatus*, bites a leaf of cassava, *Manihot esculenta*, it immediately backs away from the damage and, within a second or two, jumps off the plant. This response is the result of the sudden release of hydrogen cyanide from the damaged leaf. If a grasshopper is enclosed with a growing leaf of cassava, it dies rather than eat it. But if the leaf is wilted, rather than turgid, it becomes palatable because hydrogen cyanide is not produced, or is produced only very slowly. Under natural conditions, wilting is probably produced by the activities of large numbers of grasshoppers. This species is gregarious and if a group of them is enclosed with a growing leaf they survive (see Fig. 5.12). Apparently large numbers of bites by individuals cause the leaf to wilt so that little or no hydrogen cyanide is released from the glucosides.

The most remarkable examples of insects behaving in a way that enables them to feed on otherwise unpalatable plants occur on plants that contain laticifers or resin canals in the leaves. Before they start to feed, these insects cut through the proximal supply channels of the canal system. This reduces the flow of latex or resin to the more distal parts of the leaf and the insect is able to feed.

The pattern of cutting is adapted to the distribution pattern of the canals. In plants in which the canals branch from one or a few main veins, the insects only cut the veins; in plants with an anastomosing system of canals (see Fig. 2.6e), the insects cut a trench across the whole or a part of the leaf. Vein-cutting and trenching have been observed in a variety of insects. For example, vein-cutting behavior is exhibited by larvae of the monarch butterfly, *Danaus plexippus*, which feeds on species of *Asclepias*. Trenching behavior is exhibited by the polyphagous larvae of *Trichoplusia ni* when they feed on resin- or latex-containing plants like carrot and wild lettuce. Interestingly, however, the larvae of *T.ni* only cut trenches when feeding on plants containing resin or latex. They are induced to trench on *Plantago* which does not produce latex, by applying latex to the mandibles. Presumably chemicals in the latex cause the switch in behavior. As this example shows, this specialized behavior is not restricted to insects that specialize in feeding on latex- or resin-containing plants.

This type of behavior has clearly evolved many times and has been described in katydids, beetles and sawfly larvae, as well as in numerous caterpillars.

4.2.6 Conclusions

All green plants contain an assemblage of carbohydrates, amino acids and other compounds that are nutrients for insects and that are also phagostimulatory. Although these will differ in their concentrations, and hence phagostimulatory activity, they do not offer the basis for host selection at the plant taxonomic level. They could, however, enable the insect to determine host-plant quality within the host-plant range or within a plant.

For polyphagous insects, the balance between phagostimulants and deterrents governs the extent to which a plant is eaten, or if it is eaten at all. Phagostimulation by nutrients and other chemicals is necessary to drive feeding, but probably does not influence host-plant range, and no one phagostimulant is necessary to induce feeding. Thus host-plant range is ultimately defined by the occurrence of deterrent compounds in nonhosts.

This may also be true for some oligophagous species. Grass-feeding grasshoppers, for example, are inhibited from eating plants in families other than the Poaceae by the presence of deterrents (Fig. 4.25); no phagostimulants that characterize grasses as a group have been found. Grasses appear to be eaten because they contain no deterrents for these insects. If concentrations of deterrents are high, as during the seedling stages, even grasses are rejected.

For many oligophagous and perhaps for all monophagous insect species, however, feeding appears to be driven by a chemical or group of chemicals acting as a sign stimulus coupled with a relative lack of deterrent effects in the host plants. Since most phytophagous insects are relatively host-specific (see Chapter 1), it is probably true that this type of mechanism governs the host-plant selection of a majority of insects. Host range in these insects is determined both by the occurrence of sign stimuli in the host plants and the presence of deterrent compounds in the nonhosts.

It is possible that odor close to the leaf surface will prove to be very important for host-plant selection by many species, but currently our lack of knowledge of relevant aspects of insect behavior makes it impossible to make a judgment.

How important are the characters of the leaf surface? In general, it appears to be true that physical features may prevent an insect from feeding, but such features probably do not contribute to host specificity. Leaf-surface chemistry, however, may contribute very significantly to host specificity, especially in the case of ovipositing females which generally have no contact with the leaf contents. In some cases, at least, chemicals on the leaf surface have a dominant role in host-plant selection. Surface chemistry also contributes to specificity in leaf-feeding insects. Most of the examples known so far indicate that surface chemicals have a deterrent effect on feeding, but a few cases are known where they are phagostimulatory. In no case so far investigated have they been demonstrated to be involved in specific identification of the host by a feeding insect, but since this has been shown with respect to oviposition there is no reason to suppose that it may not also be relevant to feeding. Perhaps, like the physical features of the leaf surface, they generally have a permissive role in relation to feeding. This role may be dependent on, and perhaps modified by, association with the leaf contents when the insect bites the leaf.

4.3 What happens in the field?

Although initially based on field observations, nearly all the work that forms the basis of our understanding of host-plant selection has been carried out in the laboratory. There are good reasons for this. Most insects are very difficult to observe accurately in the field, and environmental conditions cannot be controlled. But, although it is necessary to study and dissect the behavior in the laboratory to understand the mechanisms involved, we have seen that the final outcome does not necessarily equate with the sum of the parts. What happens naturally in the field? How efficient are the orientation mechanisms we have discussed when the insect is in the complex, ever-changing natural environment? Do some insects really go round tasting everything in the habitat until they find something that is palatable? We can only answer such questions by field observations which, so far, are very limited, although there are some notable exceptions.

In this section we consider the available field data with respect to the behavioral observations in the laboratory. In addition, field studies have often given insights into a variety of ecological factors that impact host-plant selection. These are discussed in Chapter 5 together with the effects of some physiological variables.

Some observations have been made on the behavior of polyphagous aphids close to plants. These studies were only possible when populations of the aphids were so high that there were hundreds in the air at any one place and time. *Aphis fabae* moved in all directions when the air was still, but moved upwind whenever the wind rose slightly. This orientation was apparently not odor-induced. The insects would often gather in loose aggregations above or on the lee side of prominent plants. They landed equally readily on host plants and some nonhosts, but cabbage plants were avoided apparently because of the reflectance properties of the leaves. Immediately after landing, each insect probed the leaf with its proboscis. Following this behavior, all the insects that had landed on a plant that was not the appropriate taxonomic host took off again. This happened within a few minutes. Some of those landing on a taxonomically appropriate host, in this case the European spindle tree, *Euonymus*, also took off, but a majority moved to the underside of the leaf and began to feed.

Several studies have examined host-finding by ovipositing female butterflies, and there are a few accounts of the behavior of leaf-eating insects. The studies have been restricted to species that are relatively large and that could be followed for a period of time so that it was possible to observe the behavior of an individual in detail as it moved from plant to plant.

All the butterflies that have been studied appear to use visual cues in locating their host plants for oviposition. If the host is visually conspicuous to humans, and we presume also to the butterflies, visual selection may be precise. On the other hand, if the plants are visually inconspicuous the insects frequently land on plants that are not hosts.

This is particularly well illustrated by a survey of the oviposition behavior of butterflies in Sweden. Some species oviposited on plants that projected above the general level of the vegetation or that were in flower during the oviposition period. These plants were regarded by the observer as visually "apparent," and this seemed to be true for the butterflies, too. These butterflies only rarely landed on nonhosts. But other species used hosts, like those described below, that were not conspicuous and are regarded as "nonapparent." These insects made frequent visits to nonhosts (Fig. 4.28a). Three species sometimes used apparent hosts and sometimes nonapparent hosts. They were never seen to visit nonhosts when using apparent hosts, but frequently did so when using nonapparent species (Fig. 4.28b).

Several more detailed studies illustrate the problems of species using nonapparent hosts. In a study of oviposition by *Battus philenor*, the pipevine swallowtail, which lays its eggs on two species of *Aristolochia*, less than 5% of landings

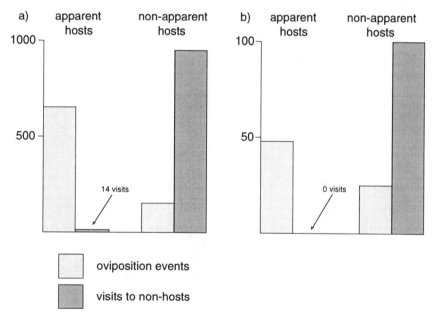

ParseFigure *4.28.* Visits by ovipositing butterflies to nonhosts in relation to the apparency of their host. "Apparent" means extending above the surrounding vegetation or in flower during the oviposition period. **a)** Based on 21 species using apparent hosts; 10 species using nonapparent hosts. Species using apparent plants for oviposition only rarely visited nonhosts. Species using nonapparent hosts made many more visits to nonhosts than oviposition visits to the host. **b)** Based on three species that sometimes used apparent and sometimes nonapparent hosts. When they were ovipositing on apparent hosts they never visited nonhosts; when they were using nonapparent plants they commonly visited nonhosts (after Wiklund, 1984).

were on the host plants. Forty-seven nonhost species from 21 families were landed on. *Eurema brigitta* is an Australian butterfly that oviposits only on *Cassia mimosoides*. Fig. 4.29 shows that only 40% of landings by ovipositing females were on the host plant, all the rest were on nonhosts. In both these instances the host plants are minor components in a complex herb layer.

This does not mean that these insects were searching and landing completely at random. *E. brigitta* tends to avoid searching in areas dominated by grasses, and other species are known to concentrate their searches in areas where the host plant is more abundant (see Chapter 5). Within these areas of search, visual recognition of host-plant characteristics further reduces the random element. Individual females of *B. philenor*, for example, tend to land either on plants with narrow leaves or on plants with broad leaves, but usually not on both types within one time period. *E. brigitta* focuses its search to an even greater extent.

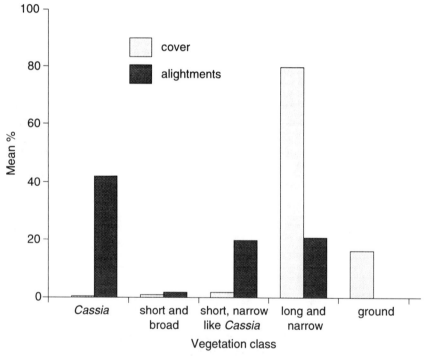

Figure 4.29. Oviposition by *Eurema brigitta*. Leaves were classified as short and narrow, like *Cassia*, the host plant, short and broad, or long and narrow. *Cassia*, and other plants with short, narrow leaves were rare in the habitat. Despite this, well over 60% of landings were made on leaves of this shape. Bars with light shading show the relative abundance of different leaf shapes in the flight paths of the butterflies. (after Mackay and Jones, 1989).

Plants with long narrow leaves, like grasses, were strongly discriminated against, but plants with short, narrow leaves, superficially resembling the host plant *Cassia*, were frequently landed on (Fig. 4.29).

Specific recognition occurs only after the butterfly has landed. In many species the female drums on the leaf with her fore tarsi. This behavior leads to acceptance of the plant and oviposition, or to its rejection. It seems evident that tarsal drumming involves contact chemoreceptors on the feet which provide information about the chemical composition of the plant, enabling the female to distinguish an acceptable from an unacceptable plant.

Landing on a taxonomically appropriate plant is not necessarily followed by oviposition. *Colias philodice*, which oviposits on a range of legume species, laid eggs on less than 50% of landings even on the most favored host (Fig. 4.30). This possibly reflects the ability of the insect to determine plant quality, so that, having landed on the right species of plant, they only laid eggs if it was in a suitable condition for larval growth and survival.

Another example where this is probably true relates to the butterfly *Perrhybris pyrrha* from Costa Rica whose caterpillars feed on a tree in the genus *Capparis*. Singer records that a female usually rejects many trees of the appropriate species before it selects one. "Then it spends up to 30 minutes apparently comparing leaves on the same tree. It flies to a leaf, alights on the tip, runs up the leaf to the base, then takes off and repeats the procedure on another leaf. Eventually, the number of leaves receiving this treatment declines from perhaps a dozen to two or three, and the insect finally vacillates between these before settling down to oviposit." Such behavior strongly suggests that the butterfly is, in some way, determining the quality of the leaves, and presumably this relates to larval survival.

An example where oviposition selection behavior has clearly been linked with larval survival concerns *B. philenor*. Its two host plants, *Aristolochia reticulata* and *A. serpentaria*, vary in their relative abundance and also in their suitability for larval survival through the season. When the plants are equally suitable for the larvae, females lay most of their eggs on the more abundant species. But when the species differ in their suitability, the females lay most of their eggs on the more suitable species even though this is the rarer of the two (Fig. 4.31). The factors used by the female in determining plant quality are not known although *B. philenor* can recognize the buds on *A. reticulata* which may provide some signal about plant quality. The buds are more common early in the season than later when the insect switches to the other host.

Landing on an appropriate host enhances the subsequent ability of the female to discriminate the host plant before landing. Species of *Colias*, which oviposit on legumes, alternate periods of oviposition with periods of nectar feeding. Immediately after visiting flowers for nectar, a female makes more errors when she resumes ovipositing, landing on nonhost plants more frequently than later

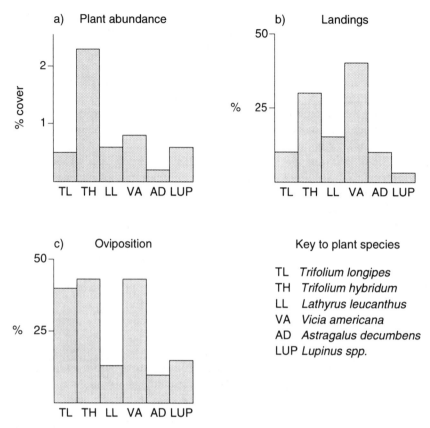

Figure 4.30. Oviposition by *Colias philodice*. **a)** Relative abundance of host plants in the habitat. These plants, all of which are legumes, only contributed a very small proportion (about 5%) to the total ground cover. **b)** Percentage of landings made on each of the plants. Landings were not determined totally by plant abundance. **c)** Percentage of landings which resulted in oviposition. Plants differed greatly in the readiness with which they were accepted (after Stanton, 1982).

in a bout of oviposition. The female clearly learns what a host species looks like, but only retains this memory for a short time because it is lost when she next undertakes a period of nectar feeding.

The oviposition behavior of *B.philenor* is similarly affected by learning. Normally a female of this species continues to search for leaves of one shape, either broad or narrow, and landing on appropriate hosts reinforces this tendency. But if, for some reason, she lands on a leaf of the opposite shape which is suitable for oviposition, she will switch to landing on leaves of this new shape.

There are very few studies in which the behavior of leaf-eating insects has been observed over long periods in the field. One concerns the horse lubber

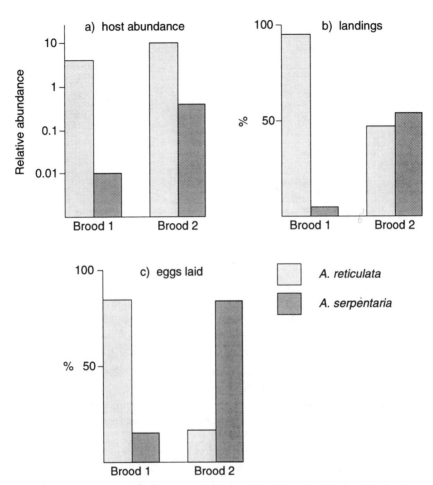

Figure 4.31. Oviposition by *Battus philenor*. **a)** Relative abundance of host plants in a suitable condition for oviposition. Note the log scale. **b)** Percentage of landings on the two hosts. **c)** Percentage of eggs on the two hosts. Early in the year (brood 1) *Aristolochia reticulata* is most suitable for larval development. Later (brood 2) *A. serpentaria* is more suitable and most eggs are laid on it despite its rarity in the habitat (after Rausher, 1980).

grasshopper, *Taeniopoda eques*. This species forages on the ground through much of the day. The longer a period of foraging, the more plant species are encountered and the more are fed on by the insects (Fig. 4.32). Most feeding bouts are of very short duration so the insect commonly takes a variety of plants before it is satiated. Wandering between bouts of feeding appears to be undirected and in the course of 30 minutes an insect might sample five or more plants and even feed on detritus on the ground (Fig. 4.33). Plants that were determined to

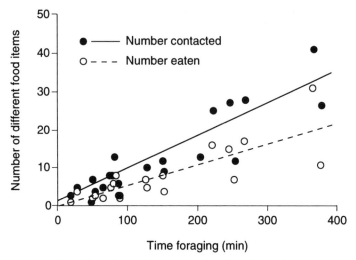

Figure 4.32. The relationship between foraging time and the number of potential food items encountered and eaten by *Taeniopoda eques*. Based on continuous observations on individuals in their natural habitat (unpublished, by courtesy of J. Howard, E.A. Bernays and D.Raubenheimer).

be highly acceptable in experiments were often rejected. There is no evidence of a directed approach to any plant, the insect simply samples each object that it encounters. If it encounters the same species more than once during a feeding period, it eats less and less on each successive feed. This is true irrespective of the acceptability of the plant (see Section 6.7).

As with oviposition, within host selection has sometimes been recorded. For example, in one series of observations, six caterpillars of the moth, *Heterocampa guttivitta*, were observed as they moved about on their host plant for five hours. During this time they walked past many leaves, and bit 98 of them. Sustained feeding only occurred on 30 of these leaves. Leaves that were rejected once continued to be rejected by the same insect on subsequent encounters and were also rejected by other individuals. Although the reasons for accepting some leaves and rejecting others are not known, the basis of choice seems likely to have been chemical.

In all the examples given above there is a strong random element in host finding. This may seem surprising, especially given the wealth of experimental evidence that insects have the ability to orient to their hosts from a distance. However, a modeling approach showed that under some circumstances random movement, coupled with arrestant responses on host plants, led to more efficient location of high quality hosts than foraging dominated by oriented responses. Interestingly, Jermy and his colleagues conclude that, despite the fact that Colo-

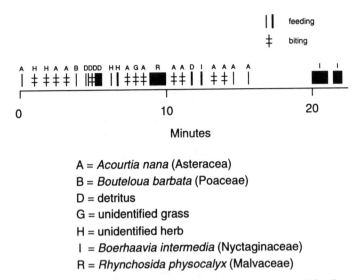

A = *Acourtia nana* (Asteracea)
B = *Bouteloua barbata* (Poaceae)
D = detritus
G = unidentified grass
H = unidentified herb
I = *Boerhaavia intermedia* (Nyctaginaceae)
R = *Rhynchosida physocalyx* (Malvaceae)

Figure 4.33. Feeding by *Taeniopoda eques*. The pattern of feeding by a single individual over a period of 22 minutes. Every plant encountered was bitten or eaten for a brief period. No feeding bout exceeded one minute (after Raubenheimer and Bernays, 1993).

rado potato beetles are attracted by the odor of a potato plant in a laboratory wind tunnel, host finding in the field is a random process.

In all these studies, the critical decisions are made when the insect arrives at the plant. Either it accepts it, or it does not. It is only possible to guess at the signals involved, although the experimental studies indicate that these will include the physical characteristics of the leaf, the chemical qualities of the surface wax and the deterrent properties of the plant sap.

It should not be supposed that these examples are representative of all insects. For example, all the butterfly studies have emphasized the role of vision, but it is likely that in other cases olfaction is more important. We might then expect the approach to a host plant to be more directed, but so far there are no comparable studies on species where this is likely to be true. Similarly, the feeding stages of many insects are less mobile than the horse lubber grasshopper, *T. eques*, and we should expect that their pattern of feeding would be quite different. There is some fragmentary information supporting this suggestion, but no extensive studies over long periods that allow detailed comparison.

Further Reading

Ahmad, S. 1983. *Herbivorous insects: host-seeking behavior and mechanisms.* Academic Press, New York.

Dethier, V.G., Barton Browne, L. and Smith, C.N. 1960. The designation of chemicals in terms of the responses they elicit from insects. J.Econ.Entomol. 53: 134–136.

Hsiao, T.H. 1985. Feeding behavior. *In* Kerkut, G.A. and Gilbert, L.I. (eds.) *Comprehensive Insect Physiology, Biochemistry and Pharmacology* vol. 9. Pergamon Press, Oxford, pp. 471–512.

Juniper, B. and Southwood, R. 1986. *Insects and the Plant Surface*. Arnold, London

Prokopy, R.J. and Owens, E.D. 1983. Visual detection of plants by herbivorous insects. A.Rev.Entomol. 28: 337–364.

Ramaswamy, S.B. (ed.) 1988. Host finding and feeding in adult phytophagous insects. J.Insect Physiol. 34: 151–268.

References (* indicates review)

Odor-induced attraction

*Bell, W.J. 1984. Chemo-orientation in walking insects. *In* Bell, W.J. and Cardé, R.T. (eds.) *Chemical Ecology of Insects*. Chapman & Hall, New York, pp. 93–109.

Bjostad, L.B. and Hibbard, B.E. 1992. 6-methoxy–2-benzoxazolinone: a semiochemical for host location by western corn rootworm larvae. J.Chem.Ecol. 18: 931–944.

*Cardé, R.T. 1984. Chemo-orientation in flying insects. *In* Bell, W.J. and Cardé, R.T. (eds.) *Chemical Ecology of Insects*. Chapman & Hall, New York, pp. 111–124.

Carle, S.A., Averill, A.L., Rule, G.S., Reissig, W.H. and Roelofs, W.L. 1987. Variation in host fruit volatiles attractive to apple maggot fly, *Rhagoletis pomonella*. J.Chem.Ecol. 13: 795–805.

Chapman, R.F., Bernays, E.A. and Simpson, S.J. 1981. Attraction and repulsion of the aphid, *Cavariella aegopodii*, by plant odors. J.Chem.Ecol. 7: 881–888.

*Elkinton, J.S. and Cardé, R.T. 1984. Odor dispersion. *In* Bell, W.J. and Cardé, R.T. (eds.) *Chemical Ecology of Insects*. Chapman & Hall, New York, pp. 73–91.

Feeny, P., Städler, E., Åhman, I. and Carter, M. 1989. effects of plant odor on oviposition by the black swallowtail butterfly, *Papilio polyxenes* (Lepidoptera: Papilionidae). J.Ins.Behav. 2: 803–827.

Fein, B.L., Reissig, W.H. and Roeloffs, W.L. 1982. Identification of apple volatiles attractive to apple maggot, *Rhagoletis pomonella*. J.Chem.Ecol. 8: 1473–1487.

Finch, S. and Skinner, G. 1982. Trapping cabbage root flies in traps baited with plant extracts and with natural and synthetic isothiocyanates. Entomologia Exp.Appl. 31: 133–139.

Hibbard, B.E. and Bjostad, L.B. 1988. Behavioral responses of western corn rootworm larvae to volatile semiochemicals from corn seedlings. J.Chem.Ecol. 14: 1523–1539.

Košťál, V. 1992. Orientation behavior of newly hatched larvae of the cabbage maggot, *Delia radicum* (L.) (Diptera: Anthomyiidae), to volatile plant metabolites. J.Insect Behav. 5: 61–70.

Lampman, R.L., Metcalf, R.L. and Abersen, J.F. 1987. Semiochemical attractants of *Diabrotica undecimpunctata howardi* Barber, southern corn rootworm, and *Diabrotica virgifera virgifera* Leconte, the western corn rootworm (Coleoptera: Chrysomelidae). J.Chem.Ecol. 13: 959–975.

Landolt, P.J. 1989. Attraction of the cabbage looper to host plants and host plant odor in the laboratory. Entomologia Exp.Appl. 53: 117–124.

Mitchell, E.R., Tingle, F.C. and Heath, R.R. 1991. Flight activity of *Heliothis virescens* (F.) females (Lepidoptera: Noctuidae) with reference to host-plant volatiles. J.Chem.Ecol. 17: 259–266.

*Murlis, J. 1986. The structure of odour plumes. *In* Payne, T.L., Birch, M.C. and Kennedy, C.E.J. (eds.) *Mechanisms in Insect Olfaction*. Clarendon Press, Oxford, pp. 27–38.

*Murlis, J., Elkinton, J.S. and Cardé, R. 1992. Odor plumes and how insects use them. A.Rev.Entomol. 37: 505–532.

Nottingham, S.F., Son, K.-C., Severson, R.F., Arrendale, R.F. and Kays, S.J. 1989. Attraction of adult sweet potato weevils, *Cylas formicarius elegantulus* (Summers), (Coleoptera: Curculionidae), to sweet potato leaf and root volatiles. J.Chem.Ecol. 15: 1095–1106.

Pivnick, K.A., Lamb, R.J. and Reed, D. 1992. Response of flea beetles, *Phyllotreta* spp., to mustard oils and nitriles in field trapping experiments. J.Chem.Ecol. 18: 863–873.

Puttick, G.M., Morrow, P.A. and Lequesne, P.W. 1988. *Trirhabda canadensis* (Coleoptera: Chrysomelidae) responses to plant odors. J.Chem.Ecol. 14: 1671–1686.

Roseland, C.R., Bates, M.B., Carlson, R.B. and Oseto, C.Y. 1992. Discrimination of sunflower volatiles by the red sunflower seed weevil. Entomologia Exp.Appl. 62: 99–106.

Ryan, M.F. and Guerin, P.M. 1982. Behavioural responses of the carrot fly larva, *Psila rosae*, to carrot root volatiles. Physiol.Entomol. 7: 315–324.

Saxena, K.N. and Prabha, S. 1975. Relationship between the olfactory sensilla of *Papilio demoleus* L. larvae and their orientation responses to different odours. J.Entomol. (A) 50: 119–126.

Sutherland, O.R.W. and Hillier, J.R. 1974. Olfactory response of *Costelytra zealandica* (Coleoptera: Melolonthinae) to the roots of several pasture plants. N.Z.J.Zool. 1: 365–369.

Thibout, E., Auger, J. and Lecomte, C. 1982. Host plant chemicals responsible for attraction and oviposition in *Acrolepiopsis assectella*. *In* Visser, J.H. and Minks, A.K. (eds.) *Insect-Plant Relationships*. Centre for Agricultural Publishing, Wageningen, pp. 107–115.

Tingle, F.C. and Mitchell, E.R. 1992. Attraction of *Heliothis virescens* (F.) (Lepidoptera: Noctuidae) to volatiles from extracts of cotton flowers. J.Chem.Ecol. 18: 907–914.

Tingle, F.C., Heath, R.R. and Mitchell, E.R. 1989. Flight response of *Heliothis subflexa* (Gn.) females (Lepidoptera: Noctuidae) to an attractant from groundcherry, *Physalis angulata* L. J.Chem.Ecol. 15: 221–231.

*Visser, J.H. 1986. Host odor perception in phytophagous insects. A.Rev.Entomol. 31: 121–144.

Visser, J.H. and Taanman, J.W. 1987. Odour-conditioned anemotaxis of apterous aphids (*Cryptomyzus korschelti*) in response to host plants. Physiol.Entomol. 12: 473–479.

Willis, M.A. and Arbas, E.A. 1991. Odor-modulated upwind flight of the sphinx moth, *Manduca sexta* L. J.Comp.Physiol. (A) 169: 427–440.

Shape and size

Harris, M.O. and Miller, J.R. 1984. Foliar form influences ovipositional behaviour of the onion fly. Physiol.Entomol. 9: 145–155.

Mackay, D.A. and Jones, R.E. 1989. Leaf shape and the host-finding behavior of two ovipositing monophagous butterfly species. Ecol.Entomol. 14: 423–431.

Saxena, K.N. and Khattar, P. 1977. Orientation of *Papilio demoleus* larvae in relation to size, distance, and combination pattern of visual stimuli. J.Insect Physiol. 23: 1421–1428.

Wallace, G.K. 1958. Some experiments on form perception in the nymphs of the desert locust, *Schistocerca gregaria* Forskål. J.Exp.Biol. 35: 765–775.

Color

Campbell, C.A.M. 1991. Response of *Phorodon humuli* to yellow and to green hop foliar colours. Entomologia Exp.Appl. 60: 95–99.

Harris, M.O. and Miller, J.R. 1983. Color stimuli and oviposition behavior of the onion fly, *Delia antiqua* (Meigen)(Diptera: Anthomyiidae). Ann.Entomol.Soc.Am. 76: 766–771.

Prokopy, R.J. 1968. Visual responses of apple maggot flies, *Rhagoletis pomonella* (Diptera: Tephritidae): orchard studies. Entomologia Exp.Appl. 11: 403–422.

Prokopy, R.J. and Owens, E.D. 1978. Visual generalist-visual specialist phytophagous insects: host selection behaviour and application to management. Entomologia Exp.Appl. 24: 609–620.

Prokopy, R.J., Collier, R.H. and Finch, S. 1983. Leaf color used by cabbage root flies to distinguish among host plants. Science 221: 190–192.

Prokopy, R.J., Collier, R.H. and Finch, S. 1983. Visual detection of host plants by cabbage root flies. Entomologia Exp.Appl. 34: 85–89.

Scherer, C. and Kolb, G. 1987. Behavioral experiments on the visual processing of color stimuli in *Pieris brassicae* L. (Lepidoptera). J.Comp.Physiol. A 160: 645–656.

Traynier, R.M.M. 1984. Associative learning in the ovipositional behaviour of the cabbage butterfly, *Pieris rapae*. Physiol.Entomol. 9: 465–472.

Vernon, R.S. and Gillespie, D.R. 1990. Spectral responsiveness of *Frankliniella occidentalis* (Thysanoptera: Thripidae) determined by trap catches in greenhouses. Environ.Entomol. 19: 1229–1241.

Physical properties of the plant

Hoffman, G.D. and McEvoy, P.B. 1985. Mechanical limitations of feeding by meadow spittlebugs *Philaenus spumarius* (Homoptera: Cercopidae) on wild and cultivated host plants. Ecol.Entomol. 10: 415–426.

Mook, J.H. 1967. Habitat selection by *Lipara lucens* Mg. (Diptera, Chloropidae) and its survival value. Arch.Néerl.Zool. 17: 469–549.

Roessingh, P. and Städler, E. 1990. Foliar form, colour and surface characteristics influence oviposition behaviour in the cabbage root fly *Delia radicum*. Entomologia Exp.Appl. 57: 93–100.

Walters, D.S., Craig, R. and Mumma, R.O. 1989. Glandular trichome exudate is the critical factor in geranium resistance to foxglove aphid. Entomologia Exp.Appl. 53: 105–109.

Leaf odor

Boer, G. de 1991. Effect of diet experience on the ability of different larval chemosensory organs to mediate food discrimination by the tobacco hornworm, *Manduca sexta*. J.Insect Physiol. 37: 763–769.

The plant surface

Blaney, W.M. and Simmonds, M.S.J. 1985. Food selection by locusts: the role of learning in rejection behaviour. Entomologia Exp.Appl. 39: 273–278.

Eigenbrode, S.D., Espelie, K.E. and Shelton, A.M. 1991. Behavior of neonate diamond-back moth larvae (*Plutella xylostella* (L.)) on leaves and on extracted leaf waxes of resistant and susceptible cabbages. J.Chem.Ecol. 17: 1691–1704.

*Chapman, R.F. 1977. The role of the leaf surface in food selection by acridoids and other insects. Coll.Internat.C.N.R.S. 265: 134–149.

*Chapman, R.F. and Bernays, E.A. 1989. Insect behavior at the leaf surface and learning as aspects of host plant selection. Experientia 45: 215–222.

Feeny, P., Sachdev, K., Rosenberry, L. and Cater, M. 1988. Luteolin 7-*o*-(6″-*o*-malonyl)-b-D-glucoside and *trans*chlorogenic acid: oviposition stimulants for the black swallowtail butterfly. Phytochemistry 27: 3439–3448.

Loon, J.J.A.van, Blaakmeer, A., Griepink, F.C., Beek, T.A.van, Schoonhoven, L.M. and Groot, A.de. 1992. Leaf surface compounds form *Brassica oleracea* (Cruciferae) induces oviposition by *Pieris brassicae* (Lepidoptera: Pieridae). Chemoecology 3: 39–44.

Renwick, J.A.A., Radke, C.D., Sachdev-Gupta, K. and Städler, E. (1992). Leaf surface chemicals stimulating oviposition by *Pieris rapae* (Lepidoptera: Pieridae) on cabbage. Chemoecology 3: 33–38.

Rivet, M.-P. and Albert, P.J. 1990. Oviposition behavior in spruce budworm *Choristoneura fumiferana* (Clem.) (Lepidoptera: Tortricidae). J.Insect Behav. 3: 395–400.

Roessingh, P., Städler, E., Schoni, R. and Feeny, P. 1991. Tarsal chemoreceptors of the black swallowtail butterfly *Papilio polyxenes*: responses to phytochemicals from host- and non-host plants. Physiol.Entomol. 16: 485–495.

Städler, E. and Buser, H.-R. 1984. Defense chemicals in leaf surface wax synergistically stimulate oviposition by a phytophagous insect. Experientia 40: 1157–1159.

Woodhead, S. 1983. Surface chemistry of *Sorghum bicolor* and its importance in feeding by *Locusta migratoria*. Physiol.Entomol. 8: 345–352.

Phagostimulants

Bernays, E.A., Howard, J.J., Champagne, D. and Estesen, B.J. 1991. Rutin: a phagostimulant for the polyphagous acridid *Schistocerca americana*. Entomologia Exp.Appl. 60: 19–28.

Blom, F. 1978. Sensory activity and food intake: a study of input-output relationships in two phytophagous insects. Netherlands J.Zool. 28: 277–340.

Bowers, M.D. 1984. Iridoid glycosides and host-plant specificity in larvae of the buckeye butterfly, *Junonia coenia* (Nymphalidae). J.Chem.Ecol. 10: 1567–1577.

Bowers, M.D. 1991. Iridoid glycosides. *In* Rosenthal, G.A. and Berenbaum, M.R. (eds.) *Herbivores. Their Interactions with Secondary Metabolites*. Academic Press, San Diego, pp. 297–325.

Chapman, R.F. and Thomas, J.G. 1978. The numbers and distribution of sensilla on the mouthparts of Acridoidea. Acrida 7: 115–148.

Ma, W.C. 1972. Dynamics of feeding responses in *Pieris brassicae* Linn. as a function of chemosensory input: a behavioural, ultrastructural and electrophysiological study. Meded.Landbouwhog.Wageningen 72, no.11.

Matsuda, K. and Matsuo, H. 1985. A flavonoid, luteolin-7-glucoside, as well as salicin and populin, stimulating the feeding of leaf beetles attacking salicaceous plants. Appl.Entomol.Zool. 20: 305–313.

Metcalf, R.L., Rhodes, A.M., Metcalf, R.A., Ferguson, J. Metcalf, E.R. and Lu, P.-Y. 1982. Cucurbitacin contents and diabroticite (Coleoptera: Chrysomelidae) feeding upon *Cucurbita* spp. Env.Entomol. 11: 931–937.

Nayer, J.K. and Thorsteinson, A.J. 1963. Further investigations into the chemical basis of insect-host plant relationships in an oligophagous insect, *Plutella maculipennis* (Curtis) (Lepidoptera:Plutellidae). Can.J.Zool. 41: 923–929.

Nielsen, J.K., Kirkeby-Thomsen, A.H. and Petersen, M.K. 1989. Host plant recognition in monophagous weevils: specificity in feeding responses of *Ceutorhynchus constrictus* and the variable effect of sinigrin. Entomologia Exp.Appl. 53: 157–166.

Soldaat, L.L. 1991. Nutritional ecology of *Tyria jacobaea* L. Ph. D. thesis, University of Leiden.

Sperling, J.H.L. and Mitchell, B.K. 1991. A comparative study of host recognition and the sense of taste in *Leptinotarsa*. J.Exp.Biol. 157: 439–459.

Pharmacophagy

Boppré, M. Redefining "pharmacophagy". 1984. J.Chem.Ecol. 10: 1151–1154.

Boppré, M., Seibt, U. and Wickler, W. 1984. Pharmacophagy in grasshoppers? Entomologia Exp.Appl. 35: 115–117.

Nishida, R. and Fukami, H. 1990. Sequestration of distasteful compounds by some pharmacophagous insects. J.Chem.Ecol. 16: 151–164.

Schneider, D. 1987. The strange fate of pyrrolizidine alkaloids. *In* Chapman, R.F., Bernays, E.A. and Stoffolano, J.G. (eds.) *Perspectives in Chemoreception and Behavior*. Springer Verlag, New York, pp. 123–142.

Deterrents

Bartlet, E. and Williams, I.H. 1991. Factors restricting the feeding of the cabbage stem flea beetle (*Psylliodes chrysocephala*). Entomologia Exp.Appl. 60: 233–238.

Bernays, E.A. and Chapman, R.F. 1976. Deterrent chemicals as a basis of oligophagy in *Locusta migratoria* (L.). Ecol.Entomol. 2: 1–18.

*Bernays, E.A. and Chapman, R.F. 1978. Plant chemistry and acridoid feeding behaviour. *In* Harborne, J.B. (ed.) *Biochemical Aspects of Plant and Animal Coevolution*. Academic Press, London, pp. 91–141.

Blaney, W.M., Simmonds, M.S.J., Ley, S.V. and Katz, R.B. 1987. An electrophysiological and behavioural study of insect antifeedant properties of natural and synthetic drimane-related compounds. Physiol.Entomol. 12: 281–291.

Blaney, W.M., Simmonds, M.S.J., Ley, S.V. and Jones, P.S. 1988. Insect antifeedants: a behavioural and electrophysiological investigation of natural and synthetically derived clerodane diterpenoids. Entomologia Exp.Appl. 46: 267–274.

Chapman, R.F., Bernays, E.A. and Wyatt, T. 1988. Chemical aspects of host-plant specificity in three *Larrea*-feeding grasshoppers. J.Chem.Ecol. 14: 561–579.

Hsiao, T.H. 1974. Chemical influence on feeding behavior of *Leptinotarsa* beetles. *In* Barton Browne, L. (ed.) *Experimental Analysis of Insect Behaviour*. Springer-Verlag, Berlin. pp 237–248.

Hsiao, T.H. 1988. Host specificity, seasonality and bionomics of Leptinotarsa beetles. *In* Jolivet, P., Petitpierre, E. and Hsiao, T.H. (eds.) *Biology of Chrysomelidae*. Kluwer, Dordrecht. pp 581–599.

Jermy, T. 1966. Feeding inhibitors and food preference in chewing phytophagous insects. Entomologia Exp.Appl. 9: 1–12.

*Morgan, E.D. and Mandava, N.B. 1985. *Handbook of Natural Pesticides vol 6, Insect Attractants and Repellents*. CRC Press, Boca Raton.

Nayar, J.K. and Thorsteinson, A.J. 1963. Further investigations into the chemical basis of insect-host plant relationships in an oligophagous insect, *Plutella maculipennis* (Curtis) (Lepidoptera: Plutellidae). Can.J.Zool. 41: 923–929.

*Norris, D.M. 1986. Anti-feeding compounds. *In Chemistry of Plant Protection* vol. 1. Springer Verlag, Berlin, pp. 97–146.

Rees, S.B. and Harborne, J.B. 1985. The role of sesquiterpene lactones and phenolics in the chemical defence of the chicory plant. Phytochemistry 24: 2225–2231.

van Drongelen, W. 1979. Contact chemoreceptors of host plant specific chemicals in larvae of various *Yponomeuta* species (Lepidoptera). J.Comp.Physiol. 134A: 265–279.

Zalucki, M.P., Brower, L.P. and Malcolm, S.B. 1990. Oviposition by *Danaus plexippus* in relation to cardenolide content of three *Asclepias* species in southeastern U.S.A. Ecol.Entomol. 15: 231–240.

Making the unpalatable palatable

Becerra, J.X. and Venable, D.L. 1990. Rapid-terpene-bath and "squirt-gun" defense in *Bursera schlechtendalii* and the counterploy of chrysomelid beetles. Biotropica 22: 320–323.

Bernays, E.A., Chapman, R.F. et al. 1977. The relationship of *Zonocerus variegatus* with cassava, *Manihot esculenta*. Bull.Entomol.Res. 67: 391–404.

Dussourd, D.E. 1993. Foraging with finesse: caterpillar adaptations for circumventing plant defenses. *In* Stamp, N.E. and Casey, T.E. (eds.) *Caterpillars*. Chapman & Hall, New York, pp. 92–131.

Dussourd, D.E. and Denno, R.F. 1991. Deactivation of plant defense: correspondence between insect behavior and secretory canal architecture. Ecology 72: 1383–1396.

Dussourd, D.E. and Eisner, T. 1987. Vein-cutting behavior: insect counterploy to the latex defense of plants. Science 237: 898–901.

Parrott, W.L., Jenkins, J.N. and McCarty, J.C. 1983. Feeding behavior of first-stage tobacco budworm (Lepidoptera: Noctuidae) on three cotton cultivars. Ann.Entomol.Soc.Am. 76: 167–170.

Zalucki, M.P. and Brower, L.P. 1992. Survival of first instar larvae of *Danaus plexippus* (Lepidoptera: Danainae) in relation to cardiac glycoside and latex content of *Asclepias humistrata* (Asclepiadaceae). Chemoecology 3: 81–93.

What happens in the field?

Damman, H. and Feeny, P. 1988. Mechanisms and consequences of selective oviposition by the zebra swallowtail butterfly. Anim.Behav. 36: 563–573.

Jermy, T., Szentesi, À. and Horvath, J. 1988. Host plant finding in phytophagous insects: the case of the Colorado potato beetle. Entomologia Exp.Appl. 49: 83–98.

Kennedy, J.S., Booth, C.O., and Kershaw, W.J.S. 1959. Host finding by aphids in the field I. Gynoparae of *Myzus persicae* (Sulzer). Ann.Appl.Biol. 47: 410–423.

Kennedy, J.S., Booth, C.O., and Kershaw, W.J.S. 1959. Host finding by aphids in the field II. *Aphis fabae* Scop. (gynoparae) and *Brevicoryne brassicae* L.; with a re-appraisal of the role of host-finding behaviour in virus spread. Ann.Appl.Biol. 47: 424–444.

Mackay, D.A. and Jones, R.E. 1989. Leaf shape and the host-finding of two ovipositing monophagous butterfly species. Ecol.Entomol. 14: 423–431.

Morris, W.F. and Kareiva, P.M. 1991. How insect herbivores find suitable host plants: the interplay between random and nonrandom movement. *In* Bernays, E.A. (ed.) *Insect-Plant Interactions* vol. 3. CRC Press, Boca Raton, pp.175–208.

Raubenhiemer, D. and Bernays, E.A. 1993. Patterns of feeding in the polyphagous grasshopper *Taeniopoda eques*: a field study. *Anim.Behav.* 45: 153–167.

Rausher, M.D. 1980. Host abundance, juvenile survival, and oviposition preferences in *Battus philenor*. Evolution 34: 342–355.

Rausher, M.D. and Papaj, D.R. 1983. Host plant selection by *Battus philenor* butterflies: evidence for individual differences in foraging behaviour. Anim.Behav. 31: 341–347.

Schultz, J.C. 1983. Habitat selection and foraging tactics of caterpillars in heterogeneous trees. *In* Denno, R.F. and McClure, M.S. (eds.) *Variable Plants and Animals in Natural and Managed Systems*. Academic Press, New York, pp. 61–90.

Singer, M.C. 1984. Butterfly-hostplant relationships: host quality, adult choice and larval success. *In* Vane-Wright, R.I. and Ackery, P.R. (eds.) *The Biology of Butterflies*. Academic Press, London, pp. 81–88.

Stanton, M.L. 1982. Searching in a patchy environment: foodplant selection by *Colias p. eriphyle* butterflies. Ecology 63: 839–853.

Stanton, M.L. 1984. Short-term learning and the searching accuracy of egg-laying butterflies. Anim.Behav. 32: 33–40.

Wiklund, C. 1984. Egg-laying patterns in butterflies in relation to their phenology and the visual apparency and abundance of their host plant. Oecologia 63: 23–29.

5

Behavior: The Impact of Ecology and Physiology

Chapter 4 dealt with the mechanisms of host-plant selection by standard insects, yet many factors in the environment as well as physiological demands from within the insect influence the host choices actually made. The ecological factors include biotic elements such as interactions with other insects as well as abiotic elements, such as temperature and wind. The physiological factors include changing nutrient requirements during development, and changing preferences over shorter time intervals depending on physiological state. These influences may completely alter what an insect chooses as a host, and override basic preferences found under standard conditions. This chapter deals with the variety of external and internal factors that interact with the basic mechanisms of host-plant selection discussed in Chapter 4.

5.1 The hosts and the habitat

For many phytophagous insect species, specialization on a habitat type, such as forest, sea shore or rocky terrain, may produce an apparently monophagous relationship when only one food plant species is present, yet the insect may actually utilize other species across its geographical range. Restriction of host use as a result of the restriction of habitat use is well known for a number of butterflies that are restricted to open grassland or to forests, and for rainforest grasshoppers that are restricted to light gaps. One of the possible reasons for specializing on habitat type is that it reduces the amount of time spent searching for a host. In any case, many aspects of selectivity are influenced by factors within the habitat. The ecology of the host plant, including abundance, distribution and dispersion of individual plants, is of particular importance.

5.1.1 Host plants

The way in which plants grow, their arrangement and distribution in nature, influences food selection by herbivores. For example, among grasshoppers that forage largely on the ground, or low in the vegetation, taller plants may not be encountered, and, even if they are highly acceptable, they are unlikely to be eaten. This is the case with *Taeniopoda eques*, the polyphagous horse lubber grasshopper. It feeds readily on mesquite tree foliage when it is encountered, or presented in an experiment, but the main foraging time is spent on the ground so that this food is not frequently encountered even though it tends to be abundant in the habitat of *T. eques*.

Generalist species are often completely opportunistic, eating largely what is most commonly available. In such cases, food selection is dominated by the host species available at the particular time and place. There are many instances of such opportunistic feeding among grasshoppers. For example, one study of the small grasshopper, *Euchorthippus pulvinatus*, in different grassland areas of the Camargue region of France, showed that there was a marked parallel between the proportions of different food plants eaten and their availability in the habitat (Fig. 5.1). In one area, the grass *Aeluropus littoralis* was the dominant plant and was eaten most; in another, another grass, *Agropyron pungens*, was the dominant plant and was eaten most; in a third area, *Phragmites communis* (grass) and *Scirpus maritimus* (sedge) were co-dominant and were eaten in similar amounts.

Where more than one potential host-plant species occurs in a habitat, differences in relative abundance can affect selection in ways that are not always obvious. A very abundant host may be chosen when alternative, perhaps more acceptable, hosts are rare. Such may be the case with some insect species in crops. For example, the moth genus, *Heliothis* has a number of species that are often pests of cotton, but for none of them is cotton a preferred host either for feeding or oviposition. In other cases, butterflies have been shown to lay more eggs on a common host plant than would be expected from its abundance relative to other, equally acceptable, but less common plants. As shown in Chapter 6, this is partly a result of learning, in that experience of a suitable host enhances the ability of the insect to refind the same species.

In a study of the grass-feeding grasshopper, *Omocestus viridulus*, it was found that individuals grazed on the more common species of grass out of proportion to its availability. As the abundant species became relatively more abundant it was eaten disproportionately more. This pattern was most extreme when the common grass was also the most acceptable, but it also sometimes occurred where the common grass was not the most preferred in laboratory choice tests. This type of selective feeding has been referred to as frequency-dependent, and contrasts with the situation where a herbivore has a fixed preference.

In contrast to such cases where the abundant plant is chosen by a polyphagous

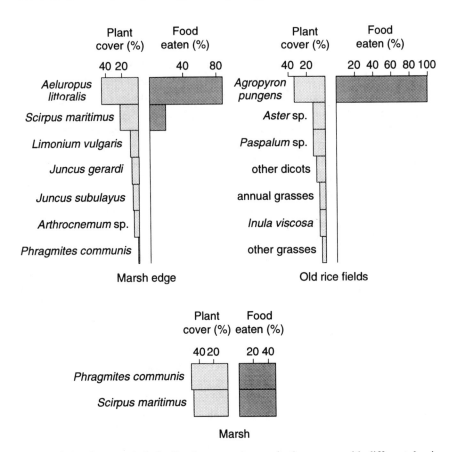

Figure 5.1. Opportunistic feeding by a grasshopper in three areas with different dominant plants. The data are from a study of the grasshopper, *Euchorthippus pulvinatus*, in the Camargue region of France. In the three different areas the proportion of the different plant species available for feeding by the grasshoppers was estimated and is presented as percent cover. The food eaten by individual grasshoppers was quantified from gut analyses. Food particles were examined microscopically and identified to species from epidermal characters, and the proportion of different foods eaten presented as a percent of the total. Most grasshoppers had only one type of food in the gut and the percent therefore generally represents the proportion of grasshoppers that had recently fed on the particular foods (after Boys, 1981).

insect more often than expected by chance, there are cases where polyphagous insects choose the rarer of the available species relatively more often than the common ones, even though these rarer species are not particularly favored in laboratory tests. If the rarer plant becomes more common it is not eaten more. Thus the proportions of different foods selected vary with their relative frequency. This type of frequency-dependent feeding was found for the desert locust, *Schis-*

tocerca gregaria, in artificial arrays of plants in the laboratory, and for the grasshopper, *Arphia sulphuria*, in field studies. Fig. 5.2 shows the results of one laboratory experiment with *S. gregaria* in which the apparent preference for two foods changed as a result of their differing proportions. As one food became rarer, it was eaten more relative to its abundance. It is important to note, however, that in this example, although the relative abundance of the plants varies, the insects maintained a consistent balance of intake between the two foods. In other words, the insects tended to eat amounts of each food that did not alter despite changes in the relative availability of the plants. It is likely that, by so doing, the insects obtained a particular balance of water or nutrients from the different food types (see Section 5.6).

A graphical representation of the different ways polyphagous species may select plants according to their relative abundance or frequencies is shown in Fig. 5.3. In one case insects may select foods opportunistically. The proportions of each plant species eaten are always similar to the proportions available. Another possibility is that they may select a plant more often than expected as its relative abundance increases. This pattern of food selection is termed pro-apostatic selection. It may imply that the favored food provides a highly suitable

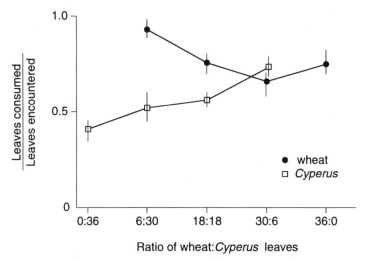

Figure 5.2. Frequency-dependent grazing by nymphs of the desert locust, *Schistocerca gregaria*, on foods available in different proportions in artificial arrays. The test involved a choice of wheat and a sedge (*Cyperus*) in differing proportions. The figure shows the proportion (mean and s.e.) of leaves eaten per encounter when wheat and *Cyperus* were available in different proportions. As a plant becomes relatively more abundant, and thus is encountered more frequently, it is also rejected more frequently. At a ratio of wheat:*Cyperus* of 30:6, each is rejected, and accepted, with similar frequency (after Chandra and Williams, 1983).

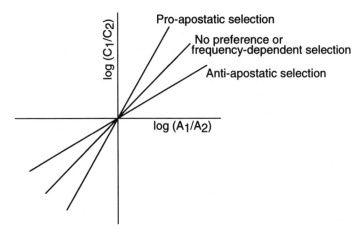

Figure 5.3. Analyses of frequency-dependent grazing by generalists. Lines represent hypothetical regression lines from frequency-dependent food selection. A_1 = number of individual plants of food 1, A_2 = number of individual plants of food type 2, C_1 = number of feeds on food 1, C_2 = number of feeds on food 2. In these examples the intercept is always = 0, indicating that the food types are of similar acceptability in general. If the slope is 1 the foods are eaten in relation to their abundance. If the slope is less than 1 the rarer food type is over-eaten. This is anti-apostatic selection. If the slope is greater than 1, the more common food type is over-eaten. This is pro-apostatic selection (after Cottam, 1985).

nutrient balance, and/or that individuals develop a preference for what they have most consistently experienced. A third possibility is that they select a food relatively more frequently as it becomes relatively less common. This pattern of food selection is termed anti-apostatic. It may indicate that individuals are actively selecting certain proportions of different foods for nutritional reasons.

Much of the work on the influence of host-plant dispersion on herbivores, has been undertaken with insect species that are relatively more host-specific than grasshoppers, or that feed on plants that are less widespread than the grasses. In these species, locating the host may provide more of a challenge, and plant dispersion will influence frequency of encounter. Thus the likelihood of finding host plants, and successfully utilizing them for oviposition or feeding, will often depend on the local density, or dispersion of the plants.

Within a plant species, it is often observed that some individuals receive eggs more consistently than others. This may be adaptive, if those receiving the eggs are better hosts for the hatching larvae. On the other hand it now seems likely that some of the variation relates simply to plant dispersion. For example, among some insect herbivores, including certain butterflies, that lay eggs on specific hosts, there is an "edge effect." That is, isolated host plants, and plants on the edge of patches of hosts, are likely to have more eggs laid on them, than plants in the middle of patches. A consequence of this effect is that smaller patches, including single plants, having a greater circumference relative to area, will have relatively more "edge" and so more eggs per plant, on average. While plant quality can be a variable associated with density, the studies on the adaptive value to the insect of such egg distribution are so far inconclusive.

A considerable amount of behavioral work has been carried out with butterflies, including the cabbage butterfly, *Pieris rapae*. Flight behavior was found to vary in relation to plant distribution. In one example, plants were either uniformly distributed or clumped. There were the same number of plants per unit area. In the uniformly distributed plot, flight lengths were shorter, and directionality greater (less turning), and there was a greater proportion of moves without displacement, compared with the plots with clumps of plants. Further observation showed that the different plant distributions also led to differences in the probability of egg deposition on plants (Fig. 5.4), there being more eggs per plant in the plot where plants were uniformly distributed than where they were clumped. In another experiment it was also shown that the more plants per plot, the fewer eggs per plant, but the more eggs laid overall in the plot.

While the behavioral parameters do not always intuitively suggest why egg distributions vary as they do, the modelling of such behaviors was found to predict very accurately both the behavior in more complex environments and the

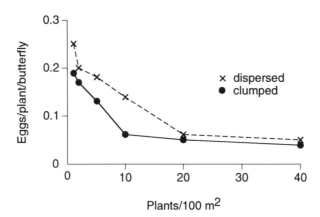

Figure 5.4. Simulated egg densities of *Pieris rapae* in relation to plant density. More eggs are laid per plant per butterfly when density of plants is low, but the pattern of change is less extreme when plants are clumped (plants in groups of four) than when they are dispersed. Similar patterns were obtained whether the dispersed plants were arranged uniformly or randomly (after Jones, 1977).

egg distributions in all the environments examined. This indicates that the measures of behavior used are relevant and while they may actually represent such things as ability to "see edges" and other unmeasurable parameters, they provide the objective measures that can explain patterns of intra-specific host selection in relation to plant dispersion.

Not all butterflies behave in the same way as *P. rapae*. In another study, with the butterfly *Cidaria albulata*, females flew shorter distances between alightments when plants were clumped than when plants were more evenly distributed. This behavior increases the chance for alightment and oviposition on the host plants in the clumps relative to the isolated plants, and thus contrasts strongly with the pattern reported for *P. rapae*.

Differences in patterns of host choice in relation to plant dispersion, seen between species of butterflies, may be related to strategies that have evolved in relation to commonly occurring distributions of host plants. But, whether this is true or not, different patches of plants may receive very different numbers of eggs, not because they differ in quality, but as a chance consequence of the female's behavior in relation to plant distribution.

The overall effect of these diverse patterns of host-plant selection is that eggs are deposited over a larger area than is necessary on the basis of host availability. The mechanisms involved include alterations in lengths of straight flights, frequency of turns, and rates of alightment, as well as factors associated with rejection of plants at close range. In terms of its functional significance, this might reduce the ability of predators and parasites to find eggs, since there is minimal information available to searching enemies about where eggs may most profitably be sought. Alternatively, the wide distribution may "spread the risks" in a general sense. In situations where juvenile survival probabilities vary markedly but unpredictably between patches, such as in disturbed habitats, the benefit is to ensure that at least some individuals survive. There is also the suggestion that a significant stochastic element in the acceptance or rejection of hosts, will tend to make sure that even the most desirable plant does not receive too many eggs.

Associated with successful oviposition or even events leading up to it, is a learning process that leads to selection of plants of the same species or variety. This has been demonstrated already in a number of butterflies, and is probably widespread among other insects. Individuals tend to select plant types they have recently oviposited on: if by chance they first encounter host A rather than host B they will then preferentially select host A. Probabilistically, females will tend to specialize on the most abundant of the available host plants at any particular time. In one study, two lycaenids were observed. *Plebejus icarioides*, which specializes on species of lupines, and *Glaucopsyche lygdamus* which utilizes several leguminous hosts, both showed highly significant preferences for the host last used for oviposition (Fig. 5.5). Furthermore, the phenologies of the butterflies

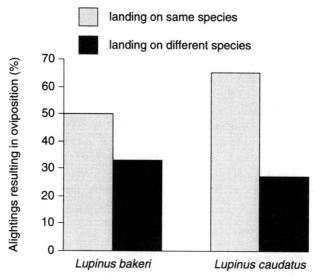

Figure 5.5. The effect of recent experience on oviposition by *Plebejus icarioides* on two species of *Lupinus*. Landing on one host species increased the likelihood that the insect would oviposit if it landed on the same species again (after Carey, 1992).

and the different host plants vary from year to year so that the insect may appear to prefer one host one year and a different host in another.

The pipevine swallowtail butterfly, *Battus philenor*, lays eggs on species of *Aristolochia*, and it was found that where two host species occur together, butterflies tended to specialize on one of them. The role of learning in this pattern is discussed in Chapter 4, but an interesting detail is that, where hosts occurred at different densities, butterflies specializing on the more abundant host discovered it more rapidly and thus laid more eggs per butterfly, than those using the less abundant host. Although it might be assumed that this was the superior strategy, offspring survival was higher among females selecting the rarer host. This highlights the fact that selection of host plants by adult Lepidoptera involves factors that relate to both adult and larval fitness.

Sometimes the more conspicuous of two equally good host species may be selected for oviposition, even though the two are present in equal densities. This has been demonstrated in butterflies that use various umbels, and which tend to select the taller plants emergent from the general vegetation.

Chrysomelid beetles seem to undertake a considerable amount of apparently random movement, independent of host patch size. Studies with crucifer flea beetles, *Phyllotreta* spp., and the cucumber specialist, *Acalymma vittata*, show that beetles move randomly within host patches, as well as in and out of host patches. This means that, as patches become smaller, emigration rates are relatively greater. It is for this reason that such beetles are uncommon in small patches, and it is rare to find them on isolated plants. In effect, the pattern of behavior of such beetles means that they select plants for feeding and oviposition that are most likely to be in a large patch. It also means that, in agricultural situations, they are relatively more abundant on their host plants in a monoculture than in a polyculture (Fig. 5.6).

In some cases, the arrangement of the vegetation influences herbivore host selection among patches of different quality. Two species of brassica flea beetle, *Phyllotreta* spp., were found to have different movement patterns dependent on the distribution of the host-plant patches. When patches were widely spaced and interpatch movements therefore low, beetles selected their food plants almost randomly with respect to patch quality. With patches closer together, there was more interpatch movement, and beetles were selective with respect to patch quality.

Very few studies have been carried out on caterpillar foraging and host selection in natural conditions in ways that can provide information on the significance of host dispersion. However, in some cases, caterpillars do have to move and make selections. It was found that *Pieris rapae* larvae that were placed in patches of hosts showed different behaviors in relation to how long they had been without food. Speed of walking and directionality increased with time (Fig. 5.7a,b). In

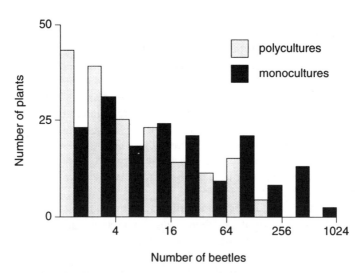

Figure 5.6. Frequency of cucumber plants with various numbers of the beetle, *Acalymma vittata*, in monocultures or in polycultures of cucumber, corn and broccoli. There were fewer beetles on the cucumber plants in polycultures apparently because the beetles moved more in the polyculture (after Bach, 1980).

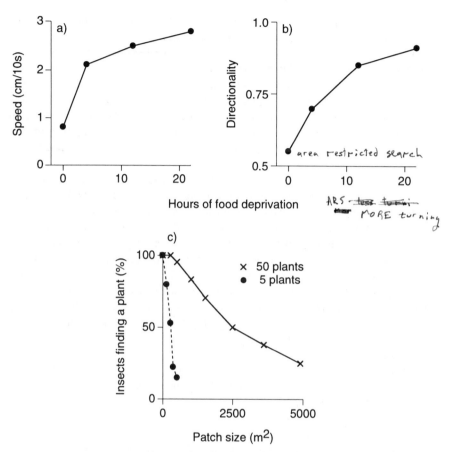

Figure 5.7. **a)** Speed and **b)** frequency of turning (directionality), in larvae of *Pieris rapae*, at increasing intervals of food deprivation at 25°C. An increase in directionality indicates a decrease in turning. **c)** Simulation of the percentages of larvae that find host plants in plots of different size having either five or 50 plants per plot. More larvae find a plant at the higher plant density. In this simulation larvae were assumed to move faster and turn less as they became more food-deprived (after Jones, 1977a).

dense stands of host plants, losing contact with a food plant is obviously not a problem since slow walking with many turns is an efficient way to refind a plant again quickly. However, if the plants are less dense and the caterpillar does not find another one soon, faster straighter movements mean that they often pass plants that are relatively close without finding them. They cover larger areas less thoroughly and move further afield. Since the nearest plants are often not found, larval populations become more spread out than might be predicted as plant density goes down. Fig. 5.7c compares host finding by larvae over 45 hours in plots of increasing size containing five or 50 plants. A smaller proportion of the plants was found at the lower of the two densities.

Other caterpillar species do not necessarily show these patterns of behavior. Another crucifer feeder, *Plutella xylostella*, does not change its behavior with increasing periods of deprivation, but continues to search in one area, covering it very thoroughly. Only plants that are close by are ever found or selected. This serves to demonstrate that the individual plants selected by larvae will often depend on the density of the host plants; plants at greater distances from the point of origin are, in practice, out of range for *P. xylostella* but not for *P. rapae*.

Temporal changes in availability of plants can also have a profound effect on what is chosen by a particular insect herbivore. It was noted above that the blue butterfly, *Glaucopsyche lygdamus*, utilizes different legume hosts depending on their phenology. In different years, host-plant species are in the appropriate stage of reproduction at different times, so host selection is dependent on the coincidence of insect flight times and flowering of the plant. There may be generational differences even within a year as shown so elegantly for *Nemoria arizonaria* on oak. This geometrid moth has two generations per year. In the spring, caterpillars of the first generation feed on catkins, upon which they are extremely cryptic; no leaves are present at this time. The second generation feeds on leaves in summer and the caterpillars are typical stem mimics. By this time there are no catkins.

There are many recorded cases of plant phenology influencing the plant chosen by phytophagous insects and the topic is discussed further in Chapter 8 with reference to host range.

5.1.2 Other plants

Neighboring plants in the community may affect host location and acceptance. In some cases the nonhost plants may make host finding difficult. For example host odors may be masked by those of nonhosts, or the odors of nonhosts may be repellent (see Chapter 4). In addition, nonhost species may reduce the visual contrast between the host and its background, or simply hide the plants from view.

Plant species other than the host may be functional requirements for some herbivores, although the extent to which precise patterns of host choice are influenced is usually not known. The simplest situation is one in which large plants may be in some way used as habitat cues. The pierid butterfly, *Euchloe belemia*, flies to patches of thorn plants. Its host plants are much more dense in the thorn patches, so that the efficiency of finding hosts is increased by concentrating on searching in thorn patches. Host plants elsewhere are ignored.

In some cases, the differing needs of larval and adult insects might determine a need for plants other than the larval host plant. For example, the celery fly, *Phylophylla heraclei*, which mines the leaves of umbels including celery, also requires trees nearby. The trees provide a mating rendezvous, and they also provide the necessary adult food in the form of homopteran honeydew (Fig. 5.8). The proximity of suitable trees may influence the use of different hosts by the flies and without them the flies are less likely to utilize the primary host.

Some adult insects require a source of nectar, and increase oviposition in the proximity of plants that provide this resource. Sometimes this need for nectar influences host choice. Populations of the butterfly, *Euphydryas editha*, in California, are more likely to use certain individual *Plantago* plants if there are additional plant species, such as *Eriodictyon*, providing nectar in the vicinity. Such constraints on host selection occur particularly in dry years when there are few plant species producing nectar during the flight periods of the butterfly. Some species of *Euphydryas* not only require a source of plants producing nectar, but also need conifers as roosting sites for males. Among heliconiine butterflies that have long lives and require pollen as a source of protein, it has been shown that *Heliconius ethilla* requires the presence of cucurbit flowers to provide the pollen. We do not know whether the appropriate availability of cucurbits for pollen actually influences the female choice of host passion vines for oviposition, but by analogy with *Euphydryas,* this is a possibility.

Sometimes, insects select nonhost plants or other objects for oviposition. A number of Lepidoptera select a substrate in the proximity of hosts, leaving the first instar larva to locate and select a plant. The noctuid moth, *Spodoptera littoralis*, often oviposits on tall perennials that are quite unsuitable as food for the larvae. The larvae hatching from the egg batches produce silk threads and

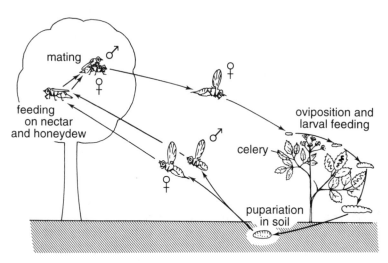

Figure 5.8. Larval celery flies are leaf miners in celery and other umbels, but trees serve as a mating rendezvous and place for adult feeding (after Leroi, 1977).

balloon down to the herbaceous annuals on which they feed. Species that feed on annuals but overwinter as eggs or small larvae are also among those that oviposit on nonhost substrates. For example, many of the fritillary butterflies, which feed on violets, first find the host plants, but then lay eggs on nonhost objects nearby. How the selection of oviposition sites is made is not known, nor how the young larvae find their hosts in the following spring.

5.2 Interactions with conspecifics and other organisms

Selection of a host plant can be influenced by the presence of other individuals of the same species, conspecifics, as well as by other herbivores, natural enemies and microorganisms. A few examples are given here to illustrate the diversity of such interactions.

5.2.1 Other insects

Conspecifics

The best example of a conspecific effect on host-plant selection is the use of oviposition marking pheromones. These pheromones are usually laid down on the substrate after the female oviposits. They have been found in Diptera, Lepidoptera, Coleoptera and Hymenoptera, and seem to predominate in flower, fruit and seed feeders. In most instances, the pheromone deters oviposition by other females of the same species, as in the apple maggot fly, *Rhagoletis pomonella*. This deterrence seems to be of value either in reducing competition or in ensuring that eggs are dispersed. In certain bruchid beetles, females deposit pheromone on the seeds after laying eggs; the number of eggs laid and the amount of pheromone deposited is related to the size of the seed. In the Mediterranean fruit fly, *Ceratitis capitata*, however, the pheromone can serve to make the fruit attractive to other females. When host fruit are large and difficult to penetrate, like oranges, a female prefers to use an oviposition hole already made where she can deposit her eggs quickly, and so, apparently, minimize the risk of predation.

Visual cues relating to conspecifics may sometimes render a plant unacceptable. Such is the case with some species of *Heliconius* butterflies which will not land if they see a conspecific egg or an egg mimic produced by the potential *Passiflora* host plant. The eggs are bright yellow, and, if they are painted green, the plant becomes acceptable. The presence of conspecific larvae sometimes deters oviposition, as it does in chalcid wasps that attack alfalfa seeds. Other examples of ovipositing insects avoiding plants with eggs or larvae of conspecifics may be found in the literature.

For the onion fly, *Delia antiqua*, it has been proposed that host compounds influence long-range attraction, but at short range, microbial products from conspecific oviposition events influence acceptance. Bacteria from females usually accompany eggs when they are laid and these bacteria were found to stimulate oviposition in laboratory tests. Final selection of oviposition site is reinforced by a marking pheromone that causes aggregation of females. Interaction of host volatiles and aggregation pheromones are best known in the many species of bark beetles (for a review see Wood, 1982). In boll weevils, some of the green leaf volatiles enhance the activity of the aggregation pheromone. This induces more insects to utilize the particular plant where activity was initiated, rather than alternative plants. A further consequence is that this could lead to more mating on such plants, making them rendezvous plants. If host plants are also rendezvous plants, then there are reproductive reasons for host fidelity.

Other species

Other herbivores can influence choice directly. For example, the cabbage root fly was deterred from laying eggs on brassica plants that had more than 250 aphids on them. They landed on the leaves but quickly took off if aphids were present, while in the absence of aphids they probed with the proboscis for some minutes then walked down the plant and laid eggs in the nearby soil.

Behavioral studies are usually lacking but there is an indication that plants may be discriminated against on the basis of prior occupation by different species of herbivore, especially when the plant is severely damaged. The level of attractiveness of potential hosts can be reduced by altering the odor, due to the frass produced and the internal volatile leaf constituents released. In addition, color changes due to early senescence caused by leaf damage, or changes in leaf shape due to galls could be important. Usually, however, the cues used by the discriminating insects are not known with certainty.

Examples of host selection being influenced by other insects most commonly involve changes in plant chemistry (see Chapter 2). In some cases, damage reduces the acceptability of the plant although this is not always true, and many of the negative effects described in the literature do not relate to behavior. One of the behavioral studies demonstrating a negative effect used larvae of the common quaker moth, *Orthosia stabilis*, that feed on silver birch. Silver birch leaves were artificially damaged in the field and the naturally infesting larvae observed on damaged and undamaged foliage. Larvae on branches with damaged foliage moved more, visited more leaves, and had more short feeding bouts on damaged than on undamaged leaves, but overall ate much more of the undamaged than the damaged leaves. This behavioral pattern indicates that the damaged leaves had become relatively unpalatable. A further example in which beetles were studied is shown in Fig. 5.9.

In some cases previous visitation by other species of herbivores can increase the acceptability of a food plant. A positive case is with two species of aphids

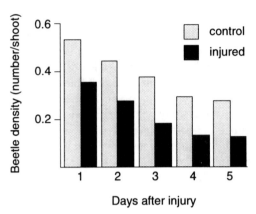

Figure 5.9.   Plant injury reduces the acceptability of a host. *Plagiodera versicolora* beetles were placed on willows with uninjured and injured shoots, on five days following the injury to the plants. On each day, more beetles were recovered from undamaged than from damaged shoots (after Raupp and Sadof, 1991).

on pine needles. Occupation of pine needles by *Schizolachnus pineti* renders them more attractive to *Eulachnus agilis* (Table 5.1).

The complexity of foraging by caterpillars in trees has been described in Chapter 4, and the behavior of individuals is suggestive of great between-leaf variation. More work is needed to discover how much of this is due to prior damage by the same or other species. In any case, the problem of finding the

Table 5.1.   *Positive interactions between two aphids on pine trees. Effect of* Schizolachnus pineti *occupation on distribution of* Eulachnus agilis. *If the* E. agilis *had been randomly distributed on the trees, the number expected to be present on needles occupied by* S. pineti *is given in the final column. In fact, the numbers observed were significantly higher on all the trees, indicating a positive interaction (after Kidd et al., 1985).*

Tree	Number of needles	Number of needles with *S. pineti*	Number of *E. agilis* on needles occupied by *S. pineti*	
			Observed	Expected
1	312	34	11	1.8
2	199	33	13	2.8
3	351	54	10	2.9
4	248	31	10	2.7
5	334	33	13	2.0
6	160	31	13	3.3

best foliage in trees may be easier in gregarious species that produce signals or trails. In the tent caterpillar, *Malacosoma neustria*, trail pheromones are used to recruit siblings to high quality feeding sites, while in larvae of *Perga* spp. (sawflies), vibrations produced by the initial foragers attract others to the site.

5.2.2 Plant-associated microorganisms

Microorganisms can influence attraction to plants. Increased attraction to host plants due to the presence of bacteria has been shown most frequently with anthomyiid flies and perhaps reflects the fact that, in many of them, the larva ultimately feeds on tissue that is undergoing some decomposition. For example, in the seed corn maggot, *Delia platura*, which is polyphagous on germinating seeds, the females are strongly attracted to the volatiles associated with a variety of bacteria.

A variety of microorganisms on the plant surface produce volatiles of their own or influence the blend of host-plant volatiles. They may have marked effects on insects selecting host plants. For example, a variety of fungi, including species of *Penicillium* and *Aspergillus*, occur on the fruit of codlings (a kind of apple). They produce odors that cause females of the yellow peach moth, *Conogethes punctiferalis*, to accumulate and lay eggs on artificial fruits in experiments (Table 5.2).

Fungal infections of plants often cause the plant to alter its metabolism which in turn influences herbivores. Most widespread, however, are the endophytes, or fungi growing within plants, which are now known to be common symbiotes,

Table 5.2. *Effect of odors from different fungi on oviposition of the yellow peach moth. Dummy fruits impregnated with the odors collected from the different fungal species were exposed to female moths in a field cage. The number of eggs laid on odor-impregnated "fruits" is compared with the number on control "fruits" (after Honda et al., 1988).*

	Number of eggs laid	
Fungal species	Test	Control
Penicillium sp.	160 ± 32	6 ± 5
Penicillium sp.	201 ± 42	1
Cladosporium sp.	306 ± 73	0
Aspergillus sp.	137 ± 15	1
Mucor sp.	24 ± 5	4 ± 4
Endothia parasitica	267 ± 52	3 ± 1
Alternaria solani	185 ± 31	3 ± 2
Glomerata cingulata	33 ± 13	1
Botrytis cinerea	15 ± 8	17 ± 7

especially among grasses. They have marked effects on the acceptability of leaves by herbivores. For example, in grasses where endophytes are very common, most infected plants are more deterrent than uninfected ones to various caterpillars, beetles and aphids. It appears that alkaloids are often produced by endophytes in grasses and many early examples of grass alkaloids are actually products of the endophytes (see Table 2.6). Many new endophytes have recently been described in pine needles and a variety of woody angiosperms, but effects on insects are so far mainly speculative.

A number of plant diseases influence the chemistry and color of their host plants and can greatly alter their attractiveness to herbivores. For example, the leafhopper vector of tungro virus of rice, *Nephotettix virescens*, is preferentially attracted to the yellow leaves of diseased plants. In several species of leafhoppers, plants that are not normally hosts will become highly acceptable and suitable if they are infected with certain diseases. The leafhopper, *Dalbulus maidis*, is one example. It is very host specific, and normally restricted to maize, but certain plants in the daisy family (Asteraceae) become acceptable when they are infected with mycoplasma-like organisms; the insects will select these for food and grow on these "nonhosts." There are suggestions that this phenomenon is quite common, but more work is needed to identify the details, because in some reports there is no indication of whether plant choice is influenced, or whether there is differential survivorship following invasion.

5.2.3 Higher trophic level organisms

Some species of lycaenid butterfly, such as *Jalmenus evagoras*, have mutualistic relationships with ants. The larvae produce secretions fed upon by ants, and the ants protect the larvae from parasites and predators. Here the female uses cues related to the presence of ants for plant acceptance. In experimental tests some eggs were laid on plants without ants, but an overwhelming majority were laid on plants with the appropriate species of ant inhabitants. The butterflies are attracted to the plants with ants and land preferentially on them (Table 5.3).

Predatory ants, including those that inhabit plants producing extrafloral nectar, and those that are hosts for homopterans that produce honeydew, can influence the choice of hosts by certain membracid bugs. This is the case with *Publilia modesta*, which preferentially accepts plants infested with conspecifics and the ants which tend them. Few nymphs survive predation from spiders without the tending ants and, as a result, the membracids become extremely dense on host plants close to the nests of servicing ants. More commonly, ants are predators of the herbivores, and their presence may affect host-plant selection. For example, some heliconiines which have no mechanisms for appeasing ants, have been shown to avoid ovipositing on *Passiflora* species with associated ants.

Paper wasps can influence within-plant food selection of caterpillars by dis-

Table 5.3. Attraction to host plants by the lycaenid butterfly, Jalmenus evagorus. *Plants were selected by the butterfly in relation to the presence of ants that may tend the larvae after they have hatched. Percentages of approaches or landings on plants with or without ants are shown in three different experiments (after Pierce and Elgar, 1985).*

	Expt.	Ants present (%)	Ants absent (%)
Approaches	1	71	29
	2	73	27
	3	68	32
Landings	1	88	12
	2	100	0
	3	92	8

turbing them when they feed. An example is the escape behavior of larvae of *Hemileuca lucina*. These caterpillars normally feed on the upper, new leaves of *Spiraea*. They are gregarious and apparently aposematic, but are sometimes attacked by vespid wasps. In a field study on this system, in which *Polistes* sp. attacked larvae, the larvae either dropped off, fought back, or walked quickly to lower levels of the plant. Thus survivors tended to be restricted to poorer, shaded, older leaves of the plant when harassed by wasps. They grew less well.

5.3 Abiotic factors

Direct and indirect effects of abiotic factors may influence the value of the habitat, or the behavioral patterns of the insects, and hence the host choice. Although there are limited numbers of documented examples, wind and temperature are probably both important.

Very small insects are constrained in terms of host finding. In particular the wind will often dominate insect movement patterns and hence the available choices. Wind, interacting with the terrain, has been shown to influence host finding. Aphids carried by winds will drop out, for example, when the wind passes over tall objects, such as hedgerows, fences, and trees. A high density of aphids usually occurs on both sides of such objects, but especially the leeward side (Fig. 5.10), and the host plants of the aphids will be determined largely by what is available in these areas.

Since insects are poikilothermic, their behavior is to a large extent dominated by temperature. Many species exhibit behavior tending to optimize body temperature, exposing themselves to maximum radiation when air temperature is low, and moving to the shade or to a higher level above ground when it is high. These activities can affect host-plant availability. In a laboratory study of the grasshopper, *Melanoplus differentialis*, cages were oriented with respect to the

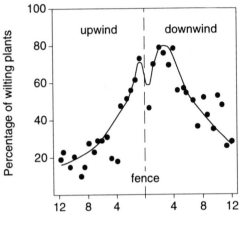

Figure 5.10. Percentage of wilting of lettuce plants following a spring migration of the aphid, *Pemphigus bursarius.* Wilting is a measure of the density of aphids on the plants. This demonstrates the way in which wind and physical factors, in this case a fence, can influence small insect distribution on hosts (after Lewis, 1965).

sun and host plants arrayed such that different species were on the sunny or shady side. Although insects preferred certain nongrasses, they ate primarily from grass species if these were on the sunny side. The effect of avoiding high temperature is illustrated by a field study of the black aposematic grasshopper, *Taeniopoda eques.* During the heat of the day this species roosts well off the ground in mesquite and acacia bushes. As the temperature drops in the afternoon, and the insect becomes active prior to descending from the roost, it often encounters edible parts of the bushes and has its first afternoon feeding bouts on them. These food items would not be part of the diet if the insects did not roost in the shrubs, and indeed, on cloudy days, when roosting does not occur, these shrubs are neither encountered nor eaten.

Caterpillars also thermoregulate behaviorally, but often their choice of food is limited to within-plant selection. In several aposematic species, feeding on the exposed, sunny, upper parts of the plant is usual, but under very hot conditions they will move to lower, shady parts of the plant, where they feed on older leaves whose nutritional quality will almost certainly be different.

Local temperature may directly influence host choice. For example, there is evidence that ovipositing females of several species of Lepidoptera fly to the sunny side of plants, or light spots in a forest. The plant part chosen is known to depend on plant chemistry, but it has also been shown in tent caterpillars in

Canada that the females oviposit on the sunny side of the tree even though there are no differences in leaf quality that influence the larval development.

In Britain, individuals of the meadow brown butterfly, *Pararge aegeria*, oviposit on a variety of grasses, but exactly which species is used depends on the temperature of the leaves, which must be between 24° and 30°C. In spring, this will be on a plant in a sunny patch, while in summer it may on a plant in shade, often in woodland. Because the available grass species are different in the areas utilized, different hosts are used.

Indirect effects of temperature or moisture can be important. Drought may differentially influence host plants and alter the acceptability ranking. It is well known that the stress of water shortage changes plant chemistry in ways that affect insects (Chapter 2).

Other elements of radiation can influence selectivity. Three species of seed chalcid wasps were shown to be selective with respect to odor and reflectance pattern only during the brightest hours of the day. Apparently, individuals orient to the appropriate host cues only when the level of polarized light exceeds a certain value.

Apart from the suitability of different parts of a plant for feeding (see Chapter 2), microclimatic factors vary considerably on any plant. The center of a large leaf is warmer than the edge, and while it is not known whether this has an effect on behavior, it is known that behavioral thermoregulation occurs and that some caterpillar species rest along the midvein. In cabbages, the distribution of caterpillars of cabbage looper, *Trichoplusia ni*, can best be explained by the short-term preference of the caterpillars for certain temperatures and humidities. This, in turn, influences the leaf ages fed upon.

There may be various interacting effects of the biotic and abiotic environment in host-plant selection. This is illustrated by a study of the rare British butterfly, *Hesperia comma*. It was found that the females preferred to lay eggs on small *Festuca ovina* plants surrounded by bare ground or scree, the plants had to be in sheltered sunny spots, there should not be too much moisture, and nectar sources were needed. Amounts of bare ground depended on amounts of grazing. The butterfly population declined following the death of rabbits from myxomatosis, because with fewer rabbits there was less grazing.

In conclusion, the study of host-plant selection in the field has revealed a number of environmental factors that may modify the choices made. In individual species of phytophagous insects, there may be several interacting factors that need to be considered for a full understanding of their behavior. This is likely to prove to be important in conservation and other applications.

5.4 Interacting behaviors

The interaction between different behaviors can influence selectivity. For example, an insect escaping from a predator must abandon feeding and perhaps the host plant. Thermoregulatory behavior may affect the availability of food as discussed in section 5.3. In Africa, many species of grasshopper roost at night, high off the ground. This is commonly interpreted as an adaptation leading to escape from predators, although it may also be related to temperature regulation. The plants chosen for roosting provide opportunities for feeding on different plants from those used during the day. For example, the red locust, *Nomadacris septemfasciata*, often roosts in tall grass which consequently comprises most of the food eaten in the morning and evening. During the day, the insects move away from the roosts into areas dominated by a sedge, and this replaces the grass as the principal food (Fig. 5.11).

The food plant available can also be influenced by sexual or territorial activity. In the grasshopper, *Ligurotettix coquilletti*, males are territorial on certain creosote bushes. The dominant individuals are able to claim nutritionally superior bushes for themselves and their mates; other individuals are thus confined to inferior foods.

Gregariousness may interact with host selection in several different ways. For example, the polyphagous grasshopper, *Zonocerus variegatus*, is able to utilize cassava partly because of its gregarious behavior. Single individuals are unable to cope with the rapid HCN release that occurs when turgid leaves of cassava are bitten, but when larger numbers are present, damage by many individuals

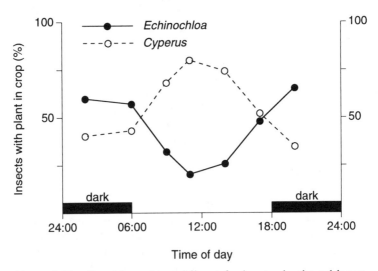

Figure 5.11. Percentage of two different foods eaten by the red locust, *Nomadacris septemfasciata*, at different times of day. The changes reflect opportunistic feeding resulting from daily movements of the insects which were unrelated to feeding. *Echinochloa* is a tall grass used for overnight roosting; *Cyperus* is a sedge growing in areas occupied by the insect for the rest of the day (after Chapman, 1957).

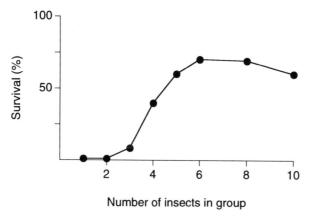

Figure 5.12. Effect of groups of insects on acceptability of food. Nymphs of the grasshopper *Zonocerus variegatus*, survive better when they are in a group on the cyanogenic crop plant cassava. Insects were enclosed on branches of growing plants in Nigeria in the field. Isolated individuals failed to survive, but survivorship increased if more than three insects were present in the enclosures (A.R. McCaffery, unpublished)

causes wilting of the plant. This in turn alters the plant chemistry, reduces HCN release rate, and allows feeding so that the insects can survive (Fig. 5.12).

5.5 Life-history constraints

Nutritional needs vary with age as they do in other animals, with protein being of particular importance in the early stages of development (Fig. 5.13). This changing need may be reflected in the changing feeding habits of insects as they grow. There are over 300 species of Lepidoptera in Britain that switch from mining to external feeding as they grow, and while this may be necessitated by increasing size, the switch also involves a big change in nutrition. Many cereal-feeding lepidopterous larvae have a pattern of diet change during development. For example, early larval instars of *Chilo partellus* feed in the young whorl leaves of the host plants where the concentration of nutrient nitrogen is relatively high. After two or three instars, larvae leave this site and crawl down to the ground. They may choose the same or a different plant but they bore into the stem, or culm, and become "stem borers," feeding on the low protein, high carbohydrate tissue. The Australian plague locust, *Chortoicetes terminifera*, is recorded as eating legumes, which are rich in protein, in the early instars, although it feeds mostly on grasses in later stages.

Sometimes ontogenetic changes in food selection may be due to morphological

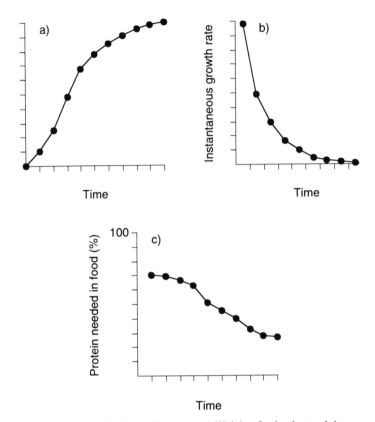

Figure 5.13. Animal growth curves. **a)** Weight of animal at each interval of time. **b)** The instantaneous growth rate for the same data. **c)** Estimated protein needs per unit time.

constraints. For example, first instar larvae of the grasshopper, *Chorthippus parallelus*, are unable to feed on the favored grass, *Festuca ovina*, because the effective leaf thickness is greater than the gape of the mandibles (Fig. 5.14). If the grass is cut into small pieces it is eaten as much or more than any other.

In many lepidopterans that feed on mature tree leaves, the early instars do not have the size or strength to feed on tough veins and are restricted to intervein areas. They have mandibles adapted to gouging out the parenchyma so that they skeletonize the leaves. Older larvae can masticate whole leaves and have mandibles adapted for snipping much like strong scissors (Fig. 5.15). This type of developmental change in feeding habit and mandible morphology has evolved in at least six different families.

The different foods used by larvae and adults of holometabolous insects are well known. In the beetle family Chrysomelidae (leaf beetles) there are many, such as the *Leptinotarsa* species, that feed on the same food plant as larvae and

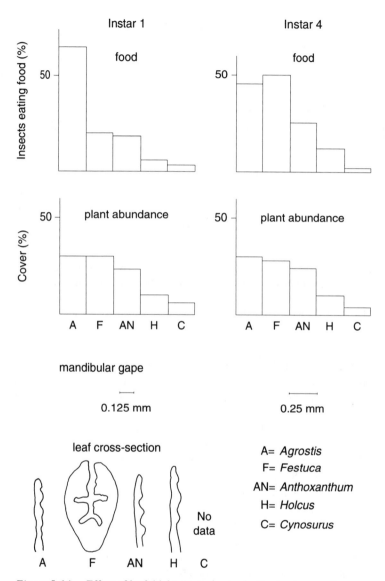

Figure 5.14. Effect of leaf thickness on food selection by the grasshopper, *Chorthippus parallelus*. In the first instar the small mandibular gape prevents most insects from feeding on *Festuca* and many more insects eat *Agrostis*. By the fourth instar the insects are big enough to feed on *Festuca* and they eat it in proportion to its availability in the habitat (after Bernays and Chapman, 1970a,b).

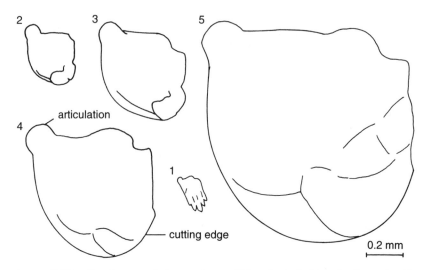

Figure 5.15. Developmental changes in mandibles of a typical tree-feeding caterpillar, *Heterocampa obliqua*, which feeds on oak. The first instar larval mandible (1) is able to penetrate the hard cuticle of oak leaves and gouge out palisade mesophyll between veins. In later instars (2-5) the mandibles are able to cut though the whole leaf including most vascular tissue (after Godfrey et al., 1989).

adults, while others make a complete change to totally unrelated host plants. For example, *Diabrotica* species feed on cereal roots as larvae and on foliage and pollen of various plants including species of Cucurbitaceae, as adults. In other coleopteran families there are other patterns of change: in several families the larvae are borers and the adults feed on pollen.

Differences in food selection occur between the sexes in some cases. *Liriomyza sativae* females feed extensively from the holes bored by the ovipositor into the leaf parenchyma. The males of this leaf-mining fly feed more from nectar. In general, females of many species of phytophagous insects do feed more, perhaps selecting for higher protein (see Section 5.6). In the tropical forest grasshopper, *Microptylopteryx hebardi*, males have a much wider host range than females, which tend to stay on plants in which they can also oviposit.

Life history constraints may influence what insects of different ages feed upon. Among Lepidoptera that undergo diapause as larvae, the plant availability and the choice made may be different before and after diapause. In the lycaenid, *Plebejus icarioides*, first instar larvae feed on old leaves of lupines in late summer and refuse to eat the younger reproductive parts of the plant. Growth is minimal. The following spring, after emergence from diapause, the larvae feed on new young growing tissue as it emerges from the ground. On this they grow fast. In a tortricid, *Gypsonoma oppressana*, the larva feeds beneath a silk web on the leaf of its food plant in autumn before the leaf is shed, and is a concealed feeder within buds the following spring. In these examples, the insects appear to select plant parts of relatively low nutrient status in the prediapause period and to reverse this selection afterwards. The effects of plant phenology in the food available to post-diapause insect is discussed in Section 8.1.

The most dramatic generational differences concern aphids. About 10% of extant aphid species show host alternation, where the sexual stages, egg and first parthenogenetic generation occur on one set of hosts, while later female generations live on a taxonomically unrelated set of hosts. The seasonal transfer is accomplished through winged migrants (Fig. 5.16). For example, *Pemphigus betae* feeds on roots of *Rumex* and *Chenopodium* species during the summer months. In October, winged migrants are produced which fly to *Populus* trees. An egg is laid under bark, and in the spring the newly hatched aphid feeds and initiates gall formation on developing leaves. The return migration is made by

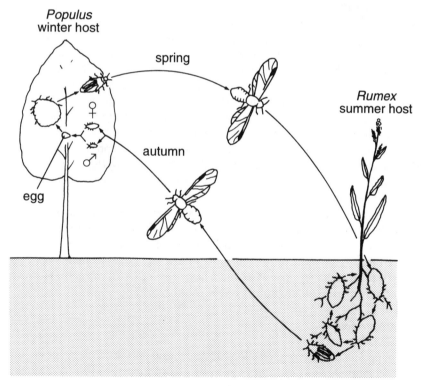

Figure 5.16. Life cycle of the aphid, *Pemphigus betae*, in Utah. Some individuals remain on roots of the summer hosts (*Rumex* or *Chenopodium* roots) throughout the winter, others migrate to *Populus angustifolia* in autumn, overwinter as sexually produced eggs and form leaf galls in spring. Individuals then migrate to herbaceous hosts in early summer (after Moran and Whitham, 1988).

parthenogenetically produced offspring which deposit nymphs beside the summer hosts. These nymphs must use cues from the plant to crawl into the soil and locate roots of the host plants (Fig. 5.16).

Less well known is the fact that obligate seasonal changes occur in host choice of ovipositing mesophyll-feeding leafhoppers. For example, females of the first summer generation of *Lindbergina aurovittata* prefer to lay eggs on oak, while the second generation chooses only blackberry. This determines larval food plants, although the larvae, when tested on the alternative food, feed on it equally well.

5.6 Physiological state variables

The physiological state of any animal influences its behavior. Which of several different behaviors is most likely to occur at any one time depends on several different known factors. The animals are responsive to certain patterns of external environmental stimuli only when specific states of the internal environment prevail. In the past the term "motivation" was often used as a definition of that class of reversible internal processes responsible for changes in behavior. For example, volumetric and osmotic changes occur if there is a shortage of water and animals will show a tendency, therefore, to drink. This is termed "thirst" and behaviorists would refer to the increased tendency to drink as a heightened "motivational state" with respect to drinking. With this terminology there are changing motivational states in relation to food, water and reproductive behavior. However, because of the everyday use of the term and the overtones of consciousness, it is preferable to refer to the physiological state variable rather than motivational state.

The level of overall excitation or arousal may have a role in selectivity. Butterflies aroused by contact with a strong oviposition stimulant in association with an unsuitable substrate will often lay an egg on a nearby nonhost. Grasshoppers artificially aroused by presentation with a highly stimulating sugar, will ingest water which they otherwise would have rejected (see Chapter 6). However, these types of arousal are probably very short-lived.

5.6.1 Nutrition

One important element of physiological state that alters selectivity relates to time since feeding. After a large meal taken by a grasshopper, the gut is distended, and endocrine changes cause inactivity. The insect moves off the food and rests, and usually remains motionless. As time passes, the insect becomes ready once again to feed, as a result of reduced gut distention, changed endocrine state, and low levels of nutrients in the blood. At this time, the greatest selectivity is shown towards potential host plants. If the insect is unable to feed, it will become progressively less selective in its choice of plant. As deprivation time increases, it accepts plants not normally eaten (Fig. 5.17).

A similar pattern has been demonstrated in many different species, and interacts with some other physiological variables. Near to the time of ecdysis, feeding rates decline, and deprivation would have less effect. Also, species that undergo reproductive diapause as adults may feed at low levels.

Adult females require more nutrients during the period of sexual maturation if they have not already developed eggs during the larval and pupal periods. This may lead to changes in feeding behavior and host-plant choice in order to obtain adequate amounts. For example in the grasshopper, *Oedaleus senega-lensis*, reproductive females choose significantly more of the protein-rich milky seed of millet, while males eat relatively more leaves (Fig. 5.18).

Nutrient imbalance can lead to changes in food selection, at least in insects feeding on artificial diets. Among both caterpillars and grasshoppers it has been demonstrated that they select different foods according to protein and carbohy-

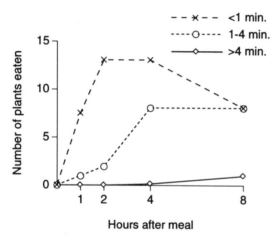

Hours after meal

Figure 5.17. Demonstration of the increasing acceptability of plants that are normally rejected after increasing periods of food deprivation in the Australian plague locust, *Chortoicetes terminifera*. Individuals were offered single foods at different times after a meal on an acceptable host. The three curves represent different feeding bout lengths. Immediately after a meal none of the 22 plants tested was acceptable. With increasing time, plants were first nibbled for less than one minute, and then eaten in moderate amounts (one to four minutes). After eight hours, one plant was eaten in relatively large amounts (>one minute) (after Bernays and Chapman, 1973).

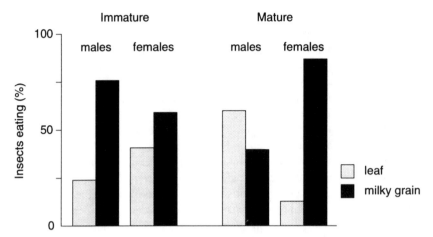

Figure 5.18. Alteration of food preferences in relation to egg development in female grasshoppers. Observations of feeding by the grasshopper, *Oedaleus senegalensis*, when offered a choice of millet leaves and young (milky) seed heads. Immature insects of both sexes preferred the grain. Mature females continued to exhibit this preference, but mature males tended to prefer the leaves (after Boys, 1978).

drate requirements. Individuals fed on diets low in carbohydrate or low in protein will, when subsequently given a choice of these two diets, select the complementary one. There is evidence that learning is involved (see Chapter 6). In the case of grasshoppers, chemoreceptor thresholds for sugars and amino acids increase as the levels increase in the blood, so that chemoreceptor input to the CNS alters. This raises the possibility that diet mixing is regulated by changes in chemosensory input.

There is very little evidence that insects feeding on natural foods select their diet to obtain complementary nutrients. However, in experiments with the grasshopper, *Schistocerca americana*, insects feeding on a mixture of plants had higher growth rates and ate less overall than did insects feeding on single plants (Fig. 5.19). In the same species, sterol deficiency, arising from feeding on certain plants with unutilizable sterols, certainly can have the effect of altering plant selection as a result of food aversion learning (see Section 6.5).

5.6.2 Water

Most phytophagous insects take in water with the food, and normally it is in adequate amounts. On the other hand most insects which have been examined have the capacity to drink, if, for one reason or another, they are suffering from a water deficit. Given a choice of foods with different moisture contents, the selection made by grasshoppers depends on their state of hydration. Insects previously fed leaves with high water content choose leaves that are drier or

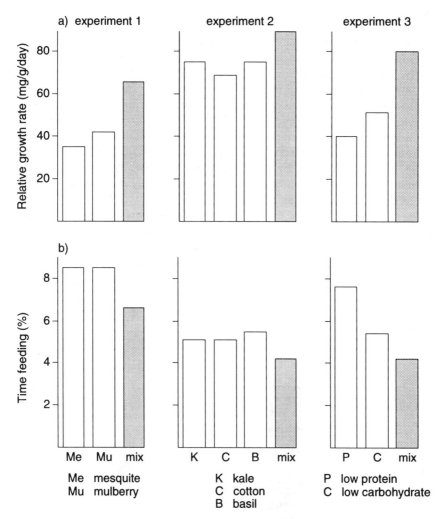

Figure 5.19. Demonstration of the value of a mixture of foods to the grasshopper, *Schistocerca americana.* **a)** Relative growth rate of sixth instar nymphs on single vs. mixed foods. **b)** Amount of feeding observed on the different diets. Insects grew at a faster rate and ate less, when they had a mixture of foods than when they had single foods. In experiment 1, insects were given mesquite or mulberry leaves or a mixture of the two; in experiment 2, insects had kale, cotton, or basil, or a mixture of all three; in experiment 3, they had an artificial diet containing low protein or low carbohydrate, or they had a choice of the two diets (Bernays and Bright, unpublished).

even totally dry, while insects previously fed on drier leaves choose leaves with high water content (Fig. 5.20). By selecting wet or dry foods grasshoppers are normally able to maintain an average water intake of 60% by weight of the food.

In the absence of free water, insects with a water deficit are tolerant of chemicals that are normally feeding deterrents. As a result, they will eat many nonhost plants, especially leaf petioles where, characteristically, the secondary metabolites are at lower levels than in leaves, and water at high levels.

Locusts generally prefer food that is relatively low in water content, but just prior to molting, water is sequestered. At that time locusts prefer wetter foods. This was shown with the migratory locust, *Locusta migratoria*, allowed to feed on fresh wheat or freeze dried wheat. During the fifth instar, approximately 65% of the food intake was the dry wheat until day eight, when the insects ate equal amounts of wet and dry food. On day nine they fed almost exclusively on the wet food, prior to molting on day ten.

Caterpillars prefer food with high water content, and given a choice of foods of the same species, they will tend to select the foliage with the highest water content. Even among aphids, where phloem feeding is apparently not constrained by water, osmotic problems are apparently a factor at times, and then they will feed instead from xylem, a very dilute fluid. Dehydrated aphids showed a higher incidence and greater duration of xylem uptake, up to 26% of a 24 hour period, compared with normally feeding individuals which feed on xylem less than 10% of the time.

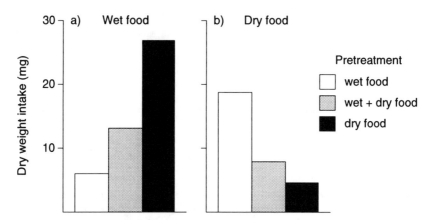

Figure 5.20. Influence of hydration on food selection by nymphs of the locust, *Schistocerca gregaria*. There were three pretreatments: Insects were fed for one day on fully hydrated wheat seedlings (wet food), or on freeze dried wheat seedlings from the same batch of seedlings (dry food), or on both, so that they could regulate their own water content. **a)** Size of the first meal when insects were tested on wet food. **b)** Size of the first meal when insects were tested on dry food. Weights are expressed as dry weight intake in all cases (after Roessingh et al., 1985).

5.6.3 Short-term feedbacks from poisons

Sometimes plant secondary metabolites are detected postingestively within minutes of being ingested. This may cause a sudden cessation of feeding activity (Fig. 5.21). Nicotine hydrogen tartrate is deterrent at initial contact when the concentration is high, but in studies with *Heliothis virescens* using lower concentrations, it was apparently not detected in the food. It reached the midgut within three minutes and apparently at that time the postingestive effects caused a sudden cessation of feeding. Such compounds have been mistakenly thought of as simple deterrents detected by the chemosensory system, and it requires close observation of behavior to determine the actual mode of action.

In some cases there is also a likelihood of aversion learning after ingestion of a poison, and this will influence food plant selection (see Section 6.5).

5.6.4 Egg load

In ovipositing females, selectivity of hosts is influenced by the egg load. Exactly what physiological factors are involved is unknown, although it has been suggested that abdominal distention is a likely influence on the behavior of the female. In any case, there are many examples of phytophagous insects becoming much less selective about their host plants when they have not recently oviposited

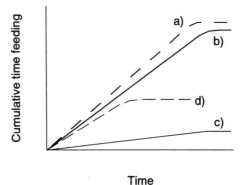

Figure 5.21. Curves illustrating different types of response by a caterpillar to deterrent chemicals in the food: **a)** expected line if caterpillar fed to repletion without stopping; **b)** a normal acceptable food on which there are short pauses; **c)** a food with a high level of deterrence. Feeding rate is low due to many pauses; **d)** food that is initially acceptable, and eaten with few pauses, but is then suddenly rejected, indicating a post-ingestive feedback.

and the egg load is large. This reduction in selectivity has been demonstrated in the butterflies, *Battus philenor* and *Euphydryas editha,* and in flies, in species of *Dacus, Rhagoletis, Delia,* and *Liriomyza.*

An increasing supply of unlaid eggs first manifests itself behaviorally in butterflies as reduced flight lengths and increased turning (Fig. 5.22). This leads, within a short time, to acceptance of plants that would be rejected when the female does not have such a big egg load. The changes in acceptability apparently occur with different thresholds of egg load or time in different individuals, and in some individuals there is no change. For example, in studies with apple maggot flies, some individuals always reject fruit that had been marked with pheromone by previous insects, while most individuals accept such fruit when the egg load becomes very high. When changes do occur, they tend to be quite abrupt (Fig. 5.23).

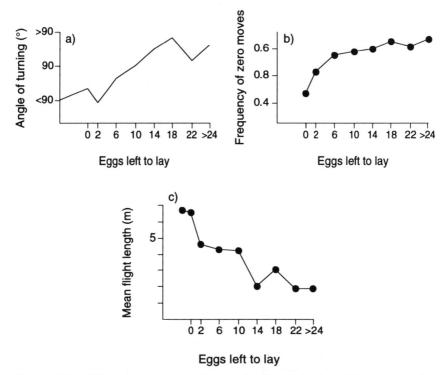

Figure 5.22. Effects of current egg load on the flight behavior of *Pieris rapae,* the cabbage butterfly. **a)** Angles of turning. **b)** Frequencies of zero moves, that is no net movement after contact with a host. **c)** Mean flight lengths. Host plants were on a square grid with 10 m between plants. As egg load increased, they turned more, returned to the same place more often, and flew shorter distances (after Jones, 1977).

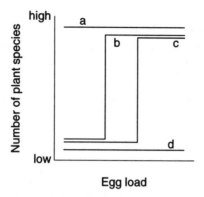

Figure 5.23. Figure to demonstrate how selectivity of host plants by herbivores decreases with egg load: **a)** is an insect type which is always relatively unselective; **b)** is one which rapidly alters with egg load; **c)** is one which requires a large egg load to change, and **d)** is one that remains selective.

5.7 Conclusions

The variables discussed in this chapter are examples of the great variety of factors, both extrinsic and intrinsic to the insect, that bear on host-plant selection. The multitude of variables modify the primary behavioral mechanisms involved. Changes may be subtle or profound, but the important point is that in understanding how insects select their plants, the mechanisms cannot be divorced from an appreciation of the additional variables. We need more detailed and continuous observations of phytophagous insects in the field to determine the role of extrinsic factors, as well as controlled laboratory experiments to put the intrinsic factors in perspective.

Further reading

Ahmad, S. (ed.) 1983. *Herbivorous Insects: Host Seeking Behavior and Mechanisms.* Academic Press, New York.

Bailey, W.J. and Ridsill-Smith, J. (eds.) 1991. *Reproductive Behaviour of Insects.* Chapman & Hall, London.

Barbosa, P., Krischik, V.A. and Jones, C.G. (eds.) 1991. *Microbial Mediation of Plant-Herbivore Interactions.* Wiley Interscience, New York.

Bernays, E.A. and Bright, K.L. 1992. Dietary mixing in grasshoppers: a review. Comp. Biochem. Physiol. A 104: 125–131

Gould, F. and Stinner, R.E. 1984. Insects in heterogeneous habitats. *In* Huffaker, C.B. and Rabb, R.L. (eds.) *Ecological Entomology.* Wiley, pp. 428–449.

Kareiva, P. 1983. Influence of vegetation texture on herbivores. *In* Denno, R.F. and McClure, M.S. (eds.) *Variable Plants and Animals in Natural and Managed Systems.* Academic Press, New York, pp. 221–289.

Miller, J.R. and Strickler, K.L. 1984. Finding and accepting host plants. *In* Bell, W. and Cardé, R. (eds.) *Chemical Ecology.* Sinauer Associates, Sunderland MA, pp. 127–157.

Montllor, C.B. 1988. The influence of plant chemistry on aphid feeding behavior. *In* Bernays, E.A. (ed.) *Insect-Plant Interactions*, vol. 3. CRC Press, Boca Raton, pp. 125–173.

Tallamy, D.W. and Raupp, M.J. 1991. *Phytochemical Induction by Herbivores.* Wiley, New York.

Thompson, J.N. and Pellmyr, O. 1991. Evolution of oviposition behavior and host preference in Lepidoptera. A.Rev.Entomol. 36: 65–90.

Wood, D.L. 1982. The role of pheromones, kairomones and allomones in the host selection and colonization behavior of bark beetles. A.Rev.Entomol. 27: 411–446.

References (* indicates review)

The hosts and the habitat

Bach,C.E. 1980. Effects of plant diversity and time of colonization on an herbivore-plant interaction. Oecologia 44: 319–326.

Boys, H.A. 1981. Food selection by some graminivorous Acrididae. D.Phil. thesis, Oxford University, U.K.

Carey, D.B. 1992. Factors determining host plant range in two lycaenid butterflies. Ph.D. thesis, University of Arizona.

Chandra, S. and Williams, G. 1983. Frequency-dependent selection in the grazing behaviour of the desert locust *Schistocerca gregaria*. Ecol.Entomol. 8: 13–21.

Cottam, D.A. 1985. Frequency-dependent grazing by slugs and grasshoppers. J.Ecol. 73: 925–933.

Courtney, S.P. 1984. Habitat vs foodplant selection. *In* Vane-Wright, R.I. and Ackery, P.R. (eds.) *The Biology of Butterflies.* Academic Press, London, pp. 89–90.

Dempster, J.P. 1982. The ecology of the cinnabar moth *Tyria jacobaeae* L. Adv.Ecol.Res. 12: 1–36.

*Ehrlich, P.R. 1984. The structure and dynamics of butterfly populations. *In* Vane-Wright, R.I. and Ackery, P.R. (eds.) *The Biology of Butterflies.* Academic Press, London, pp. 25–40.

Forsberg, J. 1987. Size discrimination among conspecific hostplants in two pierid butterflies: *Pieris napi* and *Pontia daplidice*. Oecologia 72: 52–57.

Greene, E. 1989. A diet-induced developmental polymorphism in a caterpillar. Science 243: 643–646.

Jones, R.E. 1977. Movement patterns and egg distribution in cabbage butterflies. J. Anim. Ecol. 46: 195–212.

Jones, R.E. 1977. Search behaviour: study of three caterpillar species. Behaviour 60: 237–259.

*Jones, R.E. 1991. Host location and oviposition on plants. *In* Bailey, W.J. and Ridsill-Smith, J. (eds.) *Reproductive Behaviour of Insects*. Chapman & Hall, London, pp. 108–138.

Jones, R.E., Rienks, J., Wilson, L., Lokkers, C. and Churchill, T. 1987. Temperature, development, and survival in monophagous and polyphagous tropical pierid butterflies. Aust.J.Zool. 35: 235–246.

Kareiva, P.M. 1982. Experimental and mathematical analyses of herbivore movement: quantifying the influence of plant spacing and quality on foraging discrimination. Ecol. Monogr. 52: 261–290.

Landa, K. and Rabinowitz, D. 1983. Relative preference of *Arphia sulphurea* for sparse and common prairie grasses. Ecology 64: 392–395.

Leroi, B. 1977. Biocenotic relationships of the celery fly, *Philophylla heraclei* L.: necessary presence of complementary plants for the populations living on celery. Coll.Int.CNRS, Paris, 265: 443–454.

*Maxwell, F.G., Schuster, M.F., Meredith, W.R. and Laster, M.L. 1976. Influence of the nectariless character in cotton on harmful and beneficial insects. *In* Jermy, T. (ed.) *The Host Plant in Relation to Insect Behaviour and Reproduction*. Akademiai Kaido, Budapest, pp. 157–162.

Raubenheimer, D. and Bernays, E.A. 1993. Patterns of feeding in the polyphagous grasshopper *Taeniopoda eques*: a field study. Anim.Behav.45: 153–167.

Thiery, D. and Visser, J.H. 1986. Masking of host plant odour in the olfactory orientation of the Colorado potato beetle. Entomologia Exp.Appl. 41: 165–172.

Interactions with conspecifics and other organisms

Cushman J.H. and Whitham, T.G. 1991. Competition mediating the outcome of a mutualism: protective services of ants as a limiting resource for membracids. Am.Nat. 138: 851–865.

Damman, H. and Feeny, P. 1988. Mechanisms and consequences of selective oviposition by the Zebra swallowtail Butterfly. Anim.Behav. 36: 563–573.

*Edwards, P.J., Wratten S.D. and Gibberd, R.M. 1991. The impact of inducible phytochemicals on food selection by insect herbivores and its consequences for the distribution of grazing damage. *In* Tallamy, D.W. and Raupp, M.J. (eds.) *Phytochemical Induction by Herbivores*. Wiley Interscience, New York, pp. 205–222.

Finch, S. and Jones, T.H. 1989. An analysis of the deterrent effect of aphids on cabbage root fly (*Delia radicum*) egg-laying. Ecol.Entomol. 14: 387–391.

Honda H., Ishiwatari, T. and Matsumoto, Y. 1988. Fungal volatiles as oviposition attractants for the yellow peach moth *Conogethes punctiferalis* (Guenée)(Lepidoptera: Pyralidae). J.Insect Physiol. 34: 205–212.

Hough-Goldstein, J.A. and Bassler, M.A. 1988. Effects of bacteria on oviposition by seedcorn maggots (Diptera:Anthomyiidae). Environ.Entomol. 17: 7–12.

Judd, G.J.R. and Borden J.H. 1992. Aggregated oviposition in *Delia antiqua*: a case for mediation by semiochemicals. J.Chem.Ecol. 18: 621–635.

Karban, R.1991. Inducible resistance in agricultural systems. *In* Tallamy, D.W. and Raupp, M.J. (eds.) *Phytochemical Induction by Herbivores*. Wiley Interscience, New York, pp. 403–419.

Khan, Z.R. and Saxena, R.C. 1985. Behavior and biology of *Nephotettix virescens* (Homoptera) on tungro virus-infected rice plants: epidemiology implications. Environ.Entomol. 14: 297–304.

Kidd, N.A.C., Lewis, G.B. and Howell, C.A. 1985. An association between two species of pine aphid, *Schizolachnus pineti* and *Eulachnus agilis*. Ecol.Entomol. 10: 427–432.

Peterson, S.C. 1987. Communication of leaf suitablity by gregarious eastern tent caterpillars (*Malacosoma americanum*). Ecol.Entomol. 12: 283–289.

Pierce, N.E. and Elgar, M.A. 1985. The influence of ants on host plant selection by *Jalmenus evagoras*, a myrmecophilous lycaenid butterfly. Behav.Ecol.Sociobiol. 16: 209–222.

Prokopy, R.J. and Roitberg, B.D. 1984. Foraging behavior of true fruit flies. Am.Sci. 72: 41–49.

Purcell, A.H. and Nault, L.R. 1991. Interactions among plant pathogenic prokaryotes, plants, and insect vectors. *In* Barbosa, P., Krischik, V. and Jones, C.G. (eds.) *Microbial Mediation of Plant-Herbivore Interactions*. Wiley Interscience, New York, pp. 383–405.

Raupp, M.J. and Sadof, C.S.1991. Responses of leaf beetles in injury-related changes in their salicaceous hosts. *In* Tallamy, D.W. and Raupp, M.J. (eds.) *Phytochemical Induction by Herbivores*. Wiley Interscience, New York, pp. 183–203.

Roitberg, B.D. and Mangel, M. 1988. On the evolutionary ecology of marking pheromones. Evol.Ecol. 2: 289–315.

Stamp, N.E.and Bowers, M.D. 1988. Direct and indirect effects of predatory wasps (*Polistes* sp.: Vespidae) on gregarious caterpillars (*Hemileuca lucina*: Saturniidae). Oecologia 75: 619–624.

Abiotic factors

Casey, T.M. 1992. Effects of temperature on foraging of caterpillars. *In* N.E.Stamp and T.E. Casey (eds.) *Caterpillars: Ecological and Evolutionary Constraints on Foraging* Chapman and Hall, New York, pp. 5–28.

Hoy, C.W. and Shelton, A.M. 1987. Feeding response of *Artogeia rapae* and *Trichoplusia ni* to cabbage leaf age. Environ.Entomol. 16: 680–682.

Kamm, J.A., Fairchild, C.E., Gavin, W.E. and Cooper, T.M. 1992. Influence of celestial light on visual and olfactory behavior of seed chalcids (Hymenoptera:Eurytomidae). J.Insect Behav. 5: 273–287.

Kaufmann, T. 1968. A laboratory study of feeding habits of *Melanoplus differentialis* in Maryland. Ann.Entomol.Soc.Am. 61: 173–180.

Lewis, T. 1965. The effect of an artificial windbreak on the distribution of aphids in a lettuce crop. Ann.Appl.Biol. 55:513–558.

Pedgley, D. 1982. *Windborne Pests and Diseases*. Ellis Horwood, Chichester.

Shreeve, T.G. 1986. Egg-laying by the speckled wood butterfly (*Pararge aegeria*): the role of female behaviour, host abundance and temperature. Ecol.Entomol. 11: 229–236.

Thomas, J.A., Thomas, C.D., Simcox, D.J, and Clarke, R.T. 1986. Ecology and declining status of the silver spotted skipper butterfly (*Hesperia comma*) in Britain. J.Appl Ecol. 23: 365–380.

Whitman, D.W. 1987. Thermoregulation and daily activity patterns in a black desert grasshopper, *Taeniopoda eques*. Anim.Behav. 35: 1814–1826.

Interacting behaviors

Bernays, E.A., Chapman, R.F., Leather, E.M. and McCaffery, A.R. 1977. The relationship of *Zonocerus variegatus* with cassava. Bull.Entomol.Res. 67: 391–404.

Chapman, R.F. 1957. Observations on the feeding of adults of the red locust (*Nomadacris septemfasciata*). Br.J.Anim.Behav. 2: 60–75.

Dickens, J.C. 1989. Green leaf volatiles enhance aggregation pheromone of bollweevil, *Anthonomis grandis*. Entomologia Exp. Appl. 52: 191–203.

Greenfield, M.D., Shelly, T.E. and Downum, K.R. 1987. Variation in host plant quality: implications for territoriality in a desert grasshopper. Ecology 68: 828–838.

Stanton, M.L. 1984. Short-term learning and the searching accuracy of egg-laying butterflies. Anim.Behav. 32: 33–40.

Life-history constraints

Bernays, E.A. 1992. Evolution of insect morphology in relation to plants. *In* Chaloner, W.G., Harper, J.L. and Lawton, J.H. (eds.) *The Evolutionary Interaction of Animals and Plants*. Royal Society, London, pp. 81–88.

Bernays, E.A. and Chapman, R.F. 1970a. Food selection by *Chorthippus parallelus* (Zetterstedt) (Orthoptera:Acrididae) in the field. J.Anim.Ecol. 39: 383–394.

Bernays, E.A. and Chapman, R.F. 1970b. Experiments to determine the basis of food selection by *Chorthippus parallelus* (Orthoptera: Acrididae) in the field. J.Anim.Ecol. 39: 761–776.

Bernays, E.A. and Janzen, D. 1988.Saturniid and sphingid caterpillars: two ways to eat leaves. Ecology 69: 1153–1160.

Bethke, J.A. and Parella, M.P. 1985. Leaf puncturing, feeding and oviposition behavior of *Liriomyza trifolii*. Entomologia Exp.Appl. 39: 149–154.

Braker, E. 1989. Oviposition on host plants by a tropical forest grasshopper (*Microptylopteryx hebardi*). Ecol.Entomol. 14:141–148.

Claridge, M. and Wilson, M.R. 1978. Seasonal changes and alternation of food plant preference in some mesophyll-feeding leafhoppers. Oecologia 37: 247–255.

Gaston, K.J., Reavey, D. and Valladares, G.R. 1991. Changes in feeding habit as caterpillars grow. Ecol.Entomol. 16: 339–344.

Godfrey, G.L., Miller, J.S. and Carter, D.J. 1989. Two mouthpart modifications in larval Notodontidae (Lepidoptera): their taxonomic distributions and putative functions. J.New York Entomol.Soc. 97: 455–470.

Moran, N.A. and Whitham, T.G. 1988. Evolutionary reduction of complex life cycles: loss of host-alternation in *Pemphigus*. Evolution 42: 717–728.

Physiological state variables

Bernays, E.A. 1985. Regulation of feeding behaviour. *In* Kerkut, G.A. and Gilbert, L.I. (eds.) *Comprehensive Insect Physiology, Biochemistry and Pharmacology*, vol. 4. Pergamon Press, Oxford, pp. 1–32.

Bernays, E.A. 1990. Water regulation. *In* Chapman, R.F. and Joern, A. (eds.) *Biology of Grasshoppers*. Wiley Interscience, New York, pp. 129–142.

Bernays, E.A. and Bright, K.B. 1992. Dietary mixing in grasshoppers: a review. Comp. Biochem. Physiol. A 104: 125–131.

Bernays, E.A. and Chapman, R.F. 1970. Food selection by *Chorthippus parallelus* (Orthoptera: Acrididae) in the field. J.Anim.Ecol. 39: 383–394.

Bernays, E.A. and Chapman, R.F. 1973. The role of food plants in the survival and development of *Chortoicetes terminifera* under drought conditions. Aust.J.Zool. 21: 575–592.

Bernays, E.A., Bright, K. Howard, J.J. and Raubenheimer, D. 1992. Variety is the spice of life: the basis of dietary mixing in a polyphagous grasshopper. Anim.Behav. 44: 721–731.

Bernays, E.A., Howard, J.J., Champagne, D. and Estesen, B. 1991. Rutin: a phagostimulant for the grasshopper *Schistocerca americana*. Entomologia Exp.Appl. 60: 19–28.

Boys, H.A. 1978. Food selection by *Oedaleus senegalensis* in grassland and millet fields. Entomologia Exp.Appl. 24: 278–286.

Champagne, D.and Bernays, E.A. 1991. Inadequate sterol profile as a basis of food aversion by a grasshopper. Physiol.Entomol. 16: 391–400.

Courtney, S.P. and Kibota, T.T. 1989. Mother doesn't know best: selection of hosts by ovipositing insects. *In* Bernays, E.A. (ed.) *Insect Plant Interactions*, vol. 2. CRC Press, Boca Raton, pp. 161–187.

Courtney, S., Chen, G.K. and Gardner, A. 1989. A general model for individual host selection. Oikos 55:55–65.

Jaenike, J. and Papaj, D.R. 1992. Behavioral plasticity and patterns of host use by insects. *In* B.D. Roitberg and M.B. Isman (eds.) *Insect Chemical Ecology*. Chapman and Hall, New York, pp 245–262.

Jones, R.E. 1977. Movement patterns and egg distribution in cabbage butterflies. J. Anim. Ecol. 46: 195–212.

Lewis, A.C. and Bernays, E.A. 1985. Feeding behaviour: selection of both wet and dry food for increased growth in *Schistocerca gregaria* nymphs. Entomologia Exp.Appl. 37: 105–112.

Minkenberg, O.P.J.M., Tatar, M. and Rosenheim, J.A. 1992. Egg load as a major source of variability in insect foraging and oviposition behavior. Oikos 65: 134–142.

Ng, D. 1988. A novel level of interactions in plant-insect interactions. Nature 334: 611–613.

Reynolds, S.E., Yeomans, M.R. and Timmins, W.A. 1986. The feeding behaviour of caterpillars (*Manduca sexta*) on tobacco and on artificial diet. Physiol.Entomol. 11: 39–51.

Roessingh, P., Bernays, E.A. and Lewis, A.C. 1985. Physiological factors influencing preference for wet and dry food in *Schistocerca gregaria* nymphs. Entomologia Exp. Appl. 37: 89–94.

Simpson, S.J.and Simpson, C.L. 1989. The mechanism of nutritional compensation by phytophagous insects. *In* Bernays, E.A. (ed.) *Insect Plant Interactions*, vol. 2. CRC Press, Boca Raton, pp. 111–160.

Singer, M.C. 1982. Quantification of host preference by manipulation of oviposition behaviour in the butterfly *Euphydryas editha*. Oecologia 52: 224–229.

Spiller, N.J., Koenders, L. and Tjallingii, F. 1990. Xylem ingestion by aphids—a strategy for maintaining water balance. Entomologia Exp. Appl. 55: 101–104.

6

Effects of Experience

Variation in an insect's response to a plant as a result of its previous experience is a widespread, and perhaps universal phenomenon among phytophagous insects. In some cases, the changes in host selection with experience are quite extreme, although they usually occur within the normal host range of the particular species. Several different mechanisms are involved in the changes, and at the physiological level these almost certainly overlap. The terminology is based on behavioral definitions, and comes primarily from vertebrate studies. It is not necessarily totally applicable to insects, but the parallels are important, and where possible the classical use of terms is maintained.

6.1 Definitions

Habituation: This is defined behaviorally as the waning of a response to a stimulus with repeated exposure to that stimulus. The classical example is that of an animal hearing a loud noise, which initially elicits an escape response, but with repetition is ignored. It is generally considered the simplest form of learning. It is specific to a particular stimulus and more persistent than fatigue or sensory adaptation. Thus, sensory adaptation may occur over seconds or milliseconds of exposure to a stimulus, and disadaptation occurs within minutes of the removal of the stimulus. Fatigue may last several minutes. On the other hand, habituation may last hours, days or even longer.

Physiologically, habituation has been shown in other animal groups, commonly to involve persistent synaptic changes in the specific neural pathway to the central nervous system. In short-term habituation there is a decrease in neurotransmitter release at particular synapses; in longer-term habituation there may be a decrease in productivity of neurotransmitters at synapses or even a decrease in the synaptic zone. It is possible that in some cases of habituation to chemicals there is a reduction in numbers of acceptor sites on the taste neurons.

Sensitization: This is the opposite of habituation, and involves an increased response to a stimulus on repeated exposure without any learned associations. It may occur as a result of stimuli which elicit either negative or positive behavioral effects. A classic study with crayfish demonstrates the negative aspect. Animals emerging from their crevices were confronted with very large predator images which made them retreat. On second and third repetitions of this experience, the animals retreated faster, to greater depth in the crevice, and for a longer time. In contrast to the example of habituation, where a noise with no particular significance was involved, the stimulus is a very strong negative signal.

A positive effect of sensitization is illustrated in experiments with hyenas. Individuals were presented briefly with a food item out of reach; this elicited excitatory movements and sounds. A second presentation elicited significantly more of these responses.

Several different physiological mechanisms have been proposed for sensitization. In studies with the mollusc, *Aplysia*, it results from presynaptic facilitation in interneurons. In the case of positive sensitization, some authors have proposed that it is less specific and dependant on a heightened state of excitation, although some neural pathways relevent to the experience show an increased release of neurotransmitters in synapses at the second stimulus presentation. Long-lasting sensitization involves increases in synaptic vesicles in the relevant pathways. Whatever the physiological mechanisms, sensitization is exhibited as a greater responsiveness to recently experienced biologically important stimuli.

The term "central excitatory state" has been used in many instances with respect to insect feeding behavior. The typical observation is one in which an insect contacts a strong phagostimulant and only a small quantity of food is available. If the food is removed or consumed the insect shows increased activity, turning motions, and other evidence of "searching" behavior. Although described as a central excitatory state in the entomological literature, this phenomenon is probably the same as sensitization.

Associative learning: An animal learns to associate a stimulus having no specific meaning (i.e., it is neutral), with some meaningful stimulus that produces either positive or negative effects. As a result, on subsequent encounter, the response elicited previously only by the meaningful stimulus is then elicited by the neutral stimulus. In the literature on learning, the meaningful stimulus is termed the unconditioned stimulus (US), while the neutral one that comes to be associated with it, is termed the conditioned stimulus (CS). Positive associative learning has been demonstrated in all major animal groups.

Negative associative learning has been commonly found, and in many laboratory contexts, insects have been taught to avoid certain places, behaviors or foods by punishment. This may be an electric shock or a sudden noise or a sudden bright light. In this case, the punishment becomes associated with the neutral stimulus (CS). In terms of food selection the relevant "punishments" are extreme deterrence and/or detrimental postingestive effects. The negative associations that develop between the taste of a food and subsequent illness, that have been found in numerous vertebrates, have been separately termed *aversion learning*.

Induction of preference: This term is used almost exclusively in insect studies and particularly for phytophagous insects in which individuals tend to prefer the plant they have already experienced, whether or not this plant is most appropriate for development. It has been demonstrated in different insect groups. Some authors have drawn parallels between induction and imprinting, but the classical case of imprinting described in young birds is likely to be a different phenomenon.

The physiological basis of induction is unknown, but probably involves a mixture of processes. In some cases changes in chemoreceptors are correlated with induction; there may be elements of sensitization and/or associative learning.

6.2 Habituation

Habituation to chemical deterrents is potentially a means of altering host selection behavior, enabling an insect to eat a previously unacceptable food. It has been demonstrated in a few cases. However, in some studies claiming to demonstrate habituation, this has not been distinguished from the possibility that the insects accepted the chemical more readily simply because they were short of food. This naturally occurs in experiments with deterrents, because the deterrence initially reduces feeding levels. As a result, the insects become progressively more deprived of food, and eventually may come to tolerate the deterrent (see Chapter 5 for a discussion of state variables, and their effects on host choice).

In three studies on locusts and caterpillars, habituation to deterrents has been unequivocally demonstrated. In the first case, the deterrent nicotine hydrogen tartrate was painted onto leaves of sorghum and presented to individual insects on consecutive days for 18 hours each day. In the remaining hours each day, they received uncontaminated food so that they could retain the status of well fed insects, and grow at a normal rate. On the first day of the experiment the treated leaves were deterrent to all individuals. By the third day, the nicotine-treated leaves were relatively acceptable to the desert locust, *Schistocerca gregaria*, although naive insects of similar age and similarly fed were still deterred by them. The nicotine-experienced insects had obviously habituated to this deterrent (Fig. 6.1). Interestingly, the oligophagous migratory locust, *Locusta migratoria*, showed little or no habituation over the five-day period during which it was most conspicuous with the polyphagous *S. gregaria*. Experiments with caterpillars similarly showed habituation to deterrents, and the polyphagous species habituated more than the oligophagous ones.

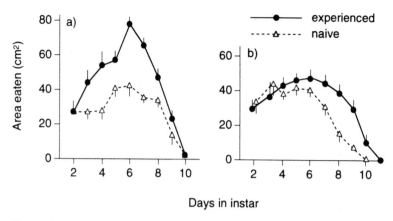

Figure 6.1. Habituation to deterrents in two species of locusts. Quantities of sorghum leaf treated with the deterrent nicotine hydrogen tartrate (NHT) consumed on different days of the final instar by **a)** *Schistocerca gregaria* and **b)** *Locusta migratoria.* Insects were either "experienced," having previously been exposed to sorghum treated with NHT, or "naive," having fed only on untreated sorghum. Experienced insects were the same individuals tested daily; naive insects were a new set of insects each day. The degree of habituation is marked in a), the polyphagous *S.gregaria*, even on the day after the start of the experiment. In b), the grass-feeding *L.migratoria*, no significant differences between the two treatments are seen until day six, and these differences may be explained by the difference in instar length, since insects having a longer instar will have greater food intake for longer. Vertical lines represent the standard errors (after Jermy et al., 1982).

Habituation to deterrents has been shown using a different experimental design with the polyphagous noctuid, *Pseudaletia unipuncta.* Individual caterpillars were exposed to deterrent caffeine-treated maize leaves, or untreated control leaves overnight. Then all individuals were allowed to take two meals on untreated corn to equalize their feeding state. In the first meal, those that had previously had the deterrent, caffeine, ate more than the controls. Presumably this was because they were relatively deprived during exposure to the deterrent and were compensating for the reduced food intake. They then had a relatively short gap before the second meal, probably for the same reason. However, by the second meal on untreated corn leaves, meal sizes were similar in both groups and it could be assumed that the treatment insects had returned to normal feeding patterns. At this time both groups were presented with caffeine-treated leaves, and the individuals which had previously experienced caffeine ate nearly twice as much as those which were naive. Clearly, they had habituated to this deterrent (Table 6.1).

Considerable variation in habituation may occur between different individuals.

Table 6.1. Habituation to a deterrent compound by larvae of Pseudaletia unipuncta.
*Lengths of meals and interfeeds taken by the larvae after overnight exposure to
caffeine-treated maize or to maize treated with solvent only. For explanation, see text
(after Usher et al., 1988).*

Larval treatment	Mean times (minutes)				
	Control meal 1	Interfeed 1	Control meal 2	Interfeed 2	Caffeine meal
Control	21	53	17	62	8
Caffeine	31	33	18	47	15
Significance: control vs caffeine	$p < 0.01$	$p < 0.05$	n.s.	n.s.	$p < 0.01$

n.s. = not significant

An example of the variation was demonstrated in experiments on the polyphagous
noctuid, *Mamestra brassicae*. Some individuals did not alter their behavior over
time, some habituated, and others apparently became sensitized (p. 221).

Although it is important to control for the feeding state of insects in the study
of habituation to deterrents, it is likely that during exposure, the insects are
driven to repeated contact with the deterrents as they search for acceptable food.
This repeated contact, enhanced by the need for food, may be important in
driving the process of habituation.

Habituation occurs to many stimuli other than those that are food related. For
example a moving shadow will elicit an escape response in a grasshopper, but
if it is repeated often, the response ceases. Such examples abound among animals.
The generality of habituation implies that it is of considerable biological significance. It is not hard to see that in the case of a moving shadow, it might signify
the approach of a predator, but if it occurs sufficiently often, it is more likely
to be caused by some waving plant. It would be a waste of time and energy to
continue to try to escape. Habituation may therefore be regarded as a mechanism
for eliminating unnecessary responses. In the case of habituation to deterrents
it is particularly relevant that a large number of deterrent compounds are harmless,
so that forfeiting a potentially good food on account of a harmless deterrent
provides a good reason for habituation. It is also logical that habituation to
deterrents would be greatest in polyphagous herbivores which are presumably
well equipped with mechanisms for dealing with plant secondary metabolites.

6.3 Sensitization

In contrast to habituation where the stimuli tend to be relatively weak, sensitization is characteristically associated with intense or highly significant stimuli such
as chemicals that are extremely deterrent or extremely phagostimulatory. A
deterrent that causes a food to be rejected after only a few seconds of feeding

may, on subsequent encounter, prevent any feeding at all. In some situations, the taste of a very deterrent compound not only increases the responsiveness to that deterrent, but in addition, leads temporarily to a reduced responsiveness to any food.

Sensitization to stimuli producing positive effects always seems to be associated with increased activity. If an insect loses contact with a highly acceptable food, it may move around extremely rapidly, giving the appearance of being "excited." Dethier first described the phenomenon with flies. He gave them access to very small drops of sugar solution or water. They usually did not drink the water upon contact, but after finding and devouring a sugar drop, the flies ran about in circles until they encountered another drop. Even if the drop was water, they would now drink it. They gave the appearance of being very excited. He considered this to result from a "central excitatory state." Later, similar experiments with hungry grasshoppers were carried out: insects were given either a drop of sucrose solution or a drop of water. They refused the water but drank the sucrose, whereupon the insects that had had sucrose were also willing to drink a drop of water. Similarly, in studies with molluscs, a highly stimulating food that caused sensitization made the animals more willing to eat something less acceptable.

A possible example of a phytophagous insect exhibiting food-related sensitization has been described for the locust, *Locusta migratoria*. Very small scattered pieces of wheat leaves were presented to individuals. An individual that encountered and fed upon one such piece became extremely active, palpated rapidly, and turned frequently until another piece was encountered. After successive encounters and subsequent ingestion, the rate of palpation and locomotor activity finally declined as the insect became satiated (Fig. 6.2).

With both habituation and sensitization it is very difficult to be sure in some instances that the processes are simple changes in the nervous system without any learned associations. For example, in the case of habituation to nicotine described above, it is possible that the insects learned to associate the deterrent with the presence of good food, and, by so learning, they chose to eat more.

What is the functional significance of sensitization? With stimuli eliciting deterrence, perhaps there is an increase in foraging efficiency since the decision not to eat may be made immediately, and the inspection of potential foods will be more rapid. On the other hand, an increasing propensity to search for and find food after a very positive experience, such as tasting a highly acceptable food, may increase the chances of obtaining more of the same without any associative learning, especially if the food occurs in patches. It is likely that the excitation is the basis of "area restricted search," discussed in some studies as an optimal approach to discovering small food items in a patch.

6.4. Associative learning or conditioning

Positive associative learning has been demonstrated with many species of phytophagous insects. In the case of oviposition, where the taste of a specific chemical

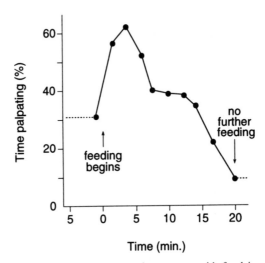

Figure 6.2. Excitation after contact with food in the migratory locust, *Locusta migratoria*. Individuals which had been deprived of food for two hours were given very small fragments of seedling wheat scattered over the substrate. After eating one piece, the insects actively turned and palpated in an apparent search behavior. The figure gives percentage of time spent palpating following loss of contact with the food at different stages in a meal. Early in the meal, very active palpation (up to 60% of time) occurs in these periods of loss of contact, illustrating excitation that probably represents sensitization (after Bernays and Chapman, 1974).

is the unconditioned stimulus, leaf color or shape may be the conditioned stimulus. For example, in the cabbage butterfly, *Pieris rapae*, females tended to prefer sites of the same appearance as those accepted previously for oviposition. Thus, following oviposition on a cabbage leaf, they increased their tendency to oviposit on substrates with a similar reflectance pattern. Sinigrin stimulates oviposition in this species, and individuals can be trained with sinigrin to oviposit on discs of different colored paper. Traynier found that, after contact with such a paper, females landed preferentially on discs of that color, even if sinigrin was absent (see Fig. 4.11, p. 117). The memory lasted for at least a day, and the butterflies could be retrained to prefer different colors or shades of green. The neutral stimulus (color) became associated with the meaningful stimulus (sinigrin).

As we have also seen in Chapter 4, *Battus philenor*, the pipevine swallowtail, selects leaves of a particular shape, depending not only on the presence of chemicals, including aristolochic acid, which apparently identify the right species, but also on the condition of the leaf. Presumably the unconditioned stimulus is a complex of chemicals and visual responses. A variety of other field observations with butterflies demonstrate that learning is widespread. The conditioned stimuli appear generally to be visual.

Where the unconditioned stimulus is the taste of food, both visual and chemical stimuli have been shown to be relevent conditioned stimuli. The grasshopper, *Melanoplus sanguinipes*, learned to forage preferentially in a place associated with a particular color/light intensity if that visual stimulus was associated with good quality food. In one experiment, individuals were observed in an arena where two differently colored boxes were provided (dark green and yellow to mimic foliage and flowers of sunflower, which are both fed upon by this species in nature). The ambient temperature was suboptimal, and a light provided warmth on the roost or resting place, so that between foraging bouts insects returned to the resting place. Naive individuals left the roost to forage and would eventually stumble on the box with the food. After feeding, they would return to the roost. The next feed was preceded by only a very short search time, even if the positions of the colored boxes were reversed, indicating an ability to learn features of the environment related to color or light intensity, and return to the food source (Table 6.2). The conditioned stimulus in this case was spectral reflectance, and the unconditioned stimulus, the food.

In a laboratory study with the locust, *Locusta migratoria*, individuals were trained to respond to specific odors associated with major essential nutrients. The locusts were trained for two days in boxes with free access to two artificial foods that were similar in all respects except that one lacked digestible carbohydrate and the other lacked protein. Each was paired with one of two distinctive odors, carvone or citral. After the training period, insects were put in clean boxes

Table 6.2.   Color learning by Melanoplus sanguinipes. *Times from the beginning of prefeeding locomotor activity on a warm roost, to the start of feeding at sites 10 centimeters away, where food was available in one of two colored boxes. The boxes were either green or yellow to mimic sunflower leaves and flowers, and in this case food was available only in the green box. Times are in minutes for individual insects in relation to the first three feeds and a subsequent feed next day. After the first feed, insects found the food much more quickly (after Bernays and Wrubel, 1985).*

Insect	Feed 1	Feed 2	Feed 3	Feed n (18–22 h later)
1	43	12	—	1
2	56	3	1	2
3	68	1	2	9
4	13	1	15	6
5	23	8	1	11
6	13	4	1	6
7	45	8	—	16
8	42	21	10	—

having no added odors and were given only one of the two diets so that they became deficient in either carbohydrate or protein. On retesting, insects deprived of carbohydrate did not respond differentially to the odors paired with carbohydrate, but those deprived of protein responded very differently to the two odors, approaching the source of the odor previously paired with the protein diet significantly more often than expected (Fig. 6.3). The experiments demonstrated that the insects had learned an association between the odors and the availability of a key nutrient they were lacking. The conditioned stimulus was the odor of the chemical, and the unconditioned stimulus, while unidentified, was presumably some nutrient feedback associated with the presence of protein. In a field situation, it is possible then, that individual grasshoppers may learn to feed on a particular food plant that best satisfies a deficit of such a major nutrient.

Clearly, grasshoppers have an ability to improve their foraging efficiency in different ways, by positive associative learning. Examples of positive associative learning have been demonstrated in other phytophagous insect groups including beetles and flies, and it seems very likely that learning abilities will be found to be widespread. However, proving associative learning is complex. Controls are required for sensitization processes (see p. 210), and appropriate experimental design for the species concerned is important since the ability to learn may be limited to ecologically relevant stimuli and situations. In a natural field situation it is often extremely difficult to prove that learning has occurred, but the profound abilities found in the laboratory imply that learning may be extremely important at least some of the time.

Apart from the direct effects of associative learning in host choices made by insects, there are implications for evolution of host-use patterns. For example, if one of two hosts is much more abundant than the other, then learning by an ovipositing insect could enhance the rate of adaptation to the abundant host. On the other hand when alternative hosts are equally abundant, learning should reinforce genetically based differences in host preference, perhaps facilitating host race formation within populations. Weight has been given to these ideas by models developed by Jaenike and Papaj. The models are based on data from some of the studies mentioned above. Further research is needed to identify exact probabilities of oviposition on various hosts and nonhosts. For example, if a butterfly has a large egg load and no available hosts, does she lay eggs on plants with only some of the suite of characteristics that are normally required? If so, and if the egg laying action provides the reward, does she learn to accept this new plant more readily, and continue to lay eggs on it, even if genuine hosts are later encountered? Could such processes precede host shifts?

6.5. Food aversion learning

Although the phenomenon of learning to reject food as a result of negative experience is obviously associative learning, it has been given special treatment

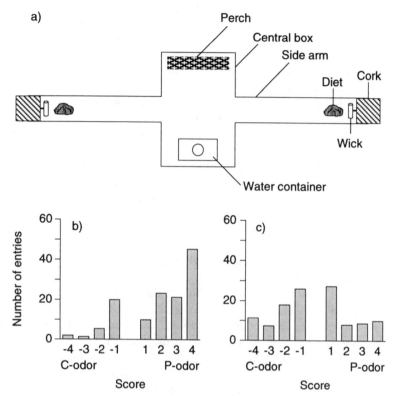

Figure 6.3. Positive associative learning of odors by *Locusta migratoria*. **a)** The experimental apparatus. A small quantity of artificial diet containing either protein, but no carbohydrate or carbohydrate, but no protein was used. The wick close to the food was dosed with a volatile compound, either carvone or citral, so that insects placed in the central box could associate a specific odor with one of the diets. During the test, the same odors were given, but both the side arms contained artificial diet lacking both carbohydrate and protein. Each insect was scored according to the distance it moved along the arms. 1 indicates that it moved into the arm, but did not move far from the central box; 4 indicates that it moved to the vicinity of the diet. **b)** Shows the results of tests with insects deprived of protein for four hours following the training period. They showed a strong tendency to move into the arm containing the odor previously associated with the protein-containing diet. **c)** Shows the results of tests with insects deprived of carbohydrate for four hours following the training period. They showed no tendency to move preferentially into either arm (after Simpson and White, 1990).

in the vertebrate literature because of the characteristic time-delay that occurs between ingestion of a food and the consequent deleterious effects. However, it has now become obvious that there is a continuum between avoidance learning unrelated to food, such as to an unacceptable physical stimulus, avoidance learning of strong deterrents, and avoidance learning of food that causes deleterious consequences upon ingestion. The latter two are of most relevance to food selection in phytophagous insects.

Clearly, an ability to learn to avoid a plant due to a noxious effect following ingestion, would be of considerable potential value. Field behavioral studies would not allow a distinction between aversion learning, sensitization, or changes in behavior brought about by other, unknown, variables. Its potential importance in food selection must initially be demonstrated in laboratory experiments, where relevant controls can ensure that the results are unambiguous.

In order to demonstrate an ability to associate a taste with a noxious effect following ingestion, nymphs of the grasshopper *Schistocerca americana*, were fed a novel food and then were injected with a poison which makes them demonstrably sick. They were then tested on the same or a different plant. It was found that individuals given novel acceptable plants after the injection ate normal-sized meals. On the other hand, individuals given the plants that they had eaten just prior to the injection ate little or none (Fig. 6.4a). This illustrates that the insects were not too sick to eat, but rejected the plant that had been eaten just before they suffered the symptoms. Additional controls showed that the chemical injected was critical, and not the injection process or handling alone. The conditioned stimulus is the taste of the first plant and the unconditioned stimulus some factor associated with the chemically-induced sickness. However, plants that were very highly acceptable in the first instance never became unacceptable. In a series of similar experiments, a variety of different secondary metabolites were shown to cause food aversion learning (Fig. 6.4b).

The probable occurrence of aversion learning with a natural food has been shown with spinach leaves. Successive meals by the grasshopper, *Schistocerca americana*, on this plant became smaller, until, after about four meals, it was rejected. Although this appeared to be food aversion learning, nonnutrients found in spinach seemed not to be involved. However, spinach contains phytosterols which are totally unsuitable for this grasshopper. Since dietary sterols are essential nutrients, there was the possibility that rejection was associated with the absence of usable sterols. Indeed, it was then found that when appropriate sterols were added to the spinach leaves, the decline in acceptability did not occur (Fig. 6.5). The sterols themselves appear not to be tasted so that the results indicate that the flavor of spinach, which was initially acceptable, became unacceptable as the insect obtained feedback concerning its nutritional unsuitability.

Aversions may also be induced by an inadequate level of protein in the food. In a series of experiments with *S. americana*, artificial diets were prepared that were either low in protein (2% wet weight) or higher in protein (4% wet weight).

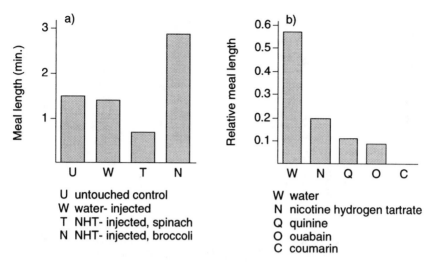

Figure 6.4. Aversion learning in the grasshopper, *Schistocerca americana.* Individual grasshoppers were given one meal on the rearing food and one on spinach. Following this, they were injected with a mild toxin, with water, or they were left untouched. Amounts injected were not usually enough to cause any obvious symptoms, although in some cases regurgitation occurred about 15 minutes after injection. **a)** The lengths of test meals on spinach or a novel food (broccoli) after injection. NHT-injected insects took only very short meals on spinach, but long meals on broccoli. This shows that the injection did not cause a general reduction in feeding, and the small meals on spinach indicate a specific association. **b)** Effects of injection of different toxins. After the injection, the insects were given a test meal on spinach. The duration of this meal is expressed as a proportion of the length of the spinach meal before injection. All four compounds tested caused a reduction in meal length (after Lee and Bernays, 1990).

Either tomatine or rutin was added at concentrations that could be detected by the grasshoppers but were not deterrent. Individuals were fed on one of the diets for four hours and then offered the low protein diet with the familiar or a novel flavor (tomatine, rutin or nontoxic levels of nicotine hydrogen tartrate). The insects that had experienced the lower protein foods fed for relatively longer periods on the diets with novel flavors than on those with the same flavor. Insects that had experienced the higher protein did not exhibit this difference. The conclusion was that insects fed protein-deficient foods subsequently showed an aversion for the flavor associated with the poor food as well as a preference for novel foods (neophilia). Similar kinds of results have often been demonstrated in studies on vertebrates, whereby learned aversions of one food go hand in hand with an increased acceptability of a novel food. In fact, in choice tests, the two cannot be separated. In any case the learned aversion for flavor coupled with low protein indicates that the neutral flavor had acquired significance, (i.e., become the conditioned stimulus).

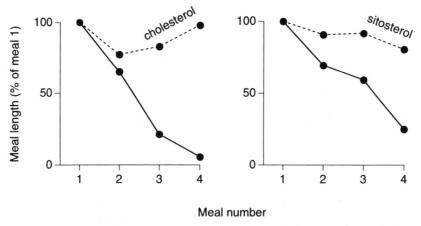

Figure 6.5. Aversion learning by grasshoppers of a food with unsuitable sterols. The solid lines show successive meal lengths by nymphs of *Schistocerca americana* on leaves of spinach. Successive meals decline in length and the fourth encounter is usually followed by rejection. When either cholesterol or sitosterol, both utilizable sterols, are added (broken lines), the learned aversion does not develop (after Champagne and Bernays, 1991).

These data indicate that aversion learning, in grasshoppers at least, may have an important role in dietary mixing. Field studies have shown considerable individual polyphagy in grasshoppers and the dietary mixing may have some basis in improving nutrient balance (see Fig. 5.19, p. 195). It is possible, for example, that successive aversion learning experiences on a series of plants that are each imperfect would lead to a better diet than remaining on one plant only. A laboratory experiment to test this idea was carried out also with *S.americana*. Two unbalanced but complementary artificial diets were offered to grasshoppers, with or without distinctive flavors, and their behavior was monitored over three days. In all cases, the grasshoppers were mobile enough to encounter and eat both diets, but the added flavors enhanced the amount of mixing. By analogy, it may be that the distinctive secondary chemistry of plants is of considerable value to phytophagous insects that are individually polyphagous, because it provides distinctive signatures and aids in the process of learning.

The evidence so far indicates that aversion learning relating to nutrient profiles causes altered preferences among different foods, and that relative acceptability of a food will vary according to both its nutrient status and the insect nutrient status/experience. *S.americana* is highly polyphagous and it is perhaps to be expected that an ability to learn from experience in this manner would be highly valuable, just as has been proposed in vertebrates. In the experiments with grasshoppers showing learned aversion to poisons, the memory of the experience was extinguished after four days. In the field this would allow sufficient time for the insect to move to a new feeding area and locate a new host plant. With nutrient deficiency, the length of the memory has not been tested, but short-term effects would be adequate to improve the efficiency of foraging.

There is one example with polyphagous arctiid caterpillars which appears to show aversion learning. *Diacrisia virginica* and *Estigmene congrua* initially respond positively to petunia leaves. After ingestion of this plant, the caterpillars regurgitated the leaf material and, when subsequently given a choice of petunia and another plant, they selected the alternative. The effect was not produced in the oligophagous caterpillar, *Manduca sexta*, even though it, too, regurgitated, and there is the suggestion that polyphagous species are better able to learn such negative associations.

6.6 Induction of preference

This phenomenon has been extensively studied among larvae of Lepidoptera where over twenty-four species have been shown to develop an altered preference in favor of the plant already experienced. An extreme example is shown in Fig. 6.6. Induction has also been demonstrated in Phasmatodea, Heteroptera, Homoptera and Coleoptera. In some extreme cases, such as the saturniid, *Callosamia promethia*, experience of one host plant apparently precluded acceptance of an alternative one.

In most published cases, induction of preference is much less extreme than the example shown in Fig. 6.6. Often, the normal host plant will remain favored. However, if larvae are forced onto an alternative plant that is accepted even though it is not a normal host, the new plant may become relatively more acceptable thereafter, but not as acceptable as the normal host. Most experiments are performed with a choice test, where relative amounts eaten are measured, so it is not easy to distinguish increased acceptability of the plant experienced from decreased acceptability of the alternative plant. However, induction of larvae of *Manduca sexta* on foods not normally eaten in nature can be explained by an increased acceptability of those plants, probably due mainly to decreases in the deterrence of certain secondary metabolites.

Further behavioral studies with artificial diets have shown that reduced deterrence of fatty acids and other lipids in a food are associated with induction on that food. Even additives such as plant secondary metabolites or artificial preservatives which are mildly deterrent have been shown to become less so in "induced" larvae.

There is evidence that there can be increased responses to stimulating extracts and odors of plants upon which larvae were induced. In experiments with *M.sexta* fed on artificial diet with or without added citral, it appeared that induction on

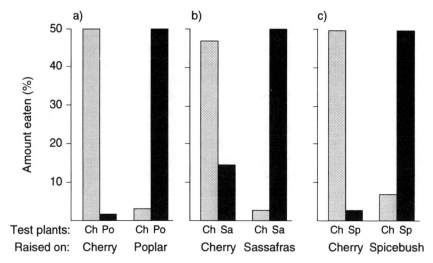

Figure 6.6. Induction of feeding preference in the larvae of *Callosamia promethia*. Results of three experiments with larvae reared and tested on different pairs of plants. **a)** Insects were raised on wild cherry (*Prunus serotina*) or tulip poplar (*Liriodendron tulipifera*) and then tested for preference between these two. **b)** Insects were raised on wild cherry or sassafras (*Sassafras albidum*) and then tested for preference between the two. **c)** Insects were raised on wild cherry or spicebush (*Lindera benzoin*) and then tested for preference between the two. Height of columns represents the percentage of each test plant eaten. Experiments were conducted in petri dishes using leaf discs with alternating plant species. The test was terminated when fifty percent of one plant species had been eaten (after Hanson, 1976).

citral-containing diet led to an ability to orient towards citral. Insects having had plain diet on the other hand, turned significantly more frequently towards the plain diet and away from citral. More realistic experiments were carried out using larvae of *Trichoplusia ni*. They were fed on either mint or basil for one week and then tested in arenas with discs of both leaf types. Not only did the larvae feed more on the plant they had experienced, as expected in an induction experiment, they also oriented to, and arrived first at, the discs of plants they had experienced previously (Table 6.3). Clearly the particular plant odors became attractants with experience. The process of induction of preference thus appears to be a mixed process phenomenon.

In examining the basis of induction, some studies have shown modifications in firing rates of chemoreceptors to various chemicals. Behavioral and electro-physiological tests with caterpillars fed on artificial diets containing a deterrent indicated that the larvae became less sensitive to the deterrent after experiencing it in the food for a period. This has been shown with *M.sexta* and salicin, and with *Spodoptera littoralis* and nicotine (Fig. 6.7). The evidence that chemoreceptor changes in sensitivity occur in relation to deterrents and phagostimulants suggests

Table 6.3. Induction of feeding preference in larvae of Trichoplusia ni. *Results of observations on larvae given a choice between discs cut from the leaves of mint and basil after individuals had fed upon one or the other, or on alfalfa for one week. Twenty individuals were used in each treatment.*

Plant previously eaten	Percentage of Insects			
	Arriving at		Taking first meal on	
	Basil	Mint	Basil	Mint
Alfalfa	65	35	60	40
Mint	5	95	0	100
Basil	80	20	95	5

that they may be part of the induction process. They do correlate with behavioral changes, but at present it is not really known if the relationship is causal. However, food clearly does alter receptor sensitivities in several different contexts. When larvae of *Manduca sexta* are reared on different plants, responses of the chemoreceptors are also different when stimulated with saps of the plants, and deterrent receptors seem to be affected more than others.

In comparing results of induction in different lepidopterous larvae, considerable variation has been found. It occurs equally readily in species with different host ranges, although the most extreme cases seem to be polyphagous. It seems that the more taxonomically different the plants in any comparison, the greater the likelihood of induction. That is, the experience of very different plants leads to a greater difference in their relative acceptability in a choice test thereafter. Another complication is that there appears to be considerable variation within species in the degree of induction that can be found in experiments. Sometimes no induction occurs even in species in which it has already been demonstrated.

Another type of induction may occur even before contact with the food plant. In adults, emergence from the pupal case may involve experience of remnants of the food of the larval stage. If such food remnants then cause a preference for those chemicals, this may be the simplest type of induction of preference. This process has been demonstrated in *Drosophila* where it has been shown that larvae reared on a flavored medium give rise to adults with a preference for this flavor over alternatives. However, if the pupae are thoroughly washed before the adult emerges this preference is eliminated. Such experiments have not been carried out with phytophagous insects, but the phenomenon may have been responsible for the earlier belief, now disfavored, that induced larval food plant preferences were transmitted to adults.

Why do so many phytophagous insects show an induction of preference? In a few laboratory experiments, it has been found that insects reared on one food will die rather than eat an alternative food which would have been accepted if it had been offered from the beginning. For example, it was shown that the cabbage butterfly, *Pieris brassicae*, could be reared on *Brassica* or on *Tropaeo-*

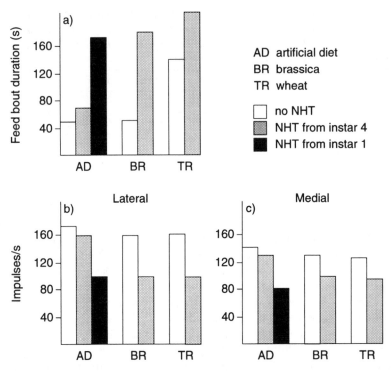

Figure 6.7. Changes in behavior and chemoreceptor responses of the noctuid, *Spodoptera littoralis,* following rearing on different diets. Larvae were reared on a bean based artificial diet, brassica or wheat. Nicotine hydrogen tartrate (NHT) was added to the diet either at instar four, or in the first instar, or not at all. **a)** The effects on behavior. Insects were tested in the final instar with filter paper containing the phagostimulant sucrose plus NHT. The histograms represent the duration of the first feeding bouts. In all cases feed duration was longer in the insects that had experienced NHT and was greatest in insects with the longer period of experience. **b), c)** The rate of firing of lateral (b) and medial (c) maxillary styloconic chemoreceptors during the first second of stimulation with an electrolyte containing 2×10^{-2}M nicotine hydrogen tartrate. Previous experience with NHT reduced the firing rates of the receptors, which are inversely correlated with feed lengths (after Blaney and Simmonds, 1987).

lum, yet if larvae were first induced on *Brassica,* they refused *Tropaeolum* and died of starvation. In such cases, induction appears to be maladaptive, but in nature this rejection behavior would lead to movement away from *Tropaeolum* if it were encountered, but not necessarily to starvation. Perhaps the significant features of induction need to be examined in the field. In any case, there has been considerable speculation as to its biological significance. In some field settings one can envisage a benefit of induction, as for example when a larva falls off its host and must find it again, a heightened sensitivity to host odors may be useful. There may be some benefit of induction during normal feeding activities on the host plant if it heightens arousal and minimizes interruptions to feeding. On the other hand, it must also be true that when the food plant is in short supply, nonpreference for an alternative would be a disadvantage.

It would be an advantage to have an induced preference if larvae became particularly adept at digesting or detoxifying the preferred plant, but evidence for this is contradictory so far. This is mainly because it is difficult to separate the effects of not feeding from the effects of reduced digestive performance.

Since ecological or physiological benefits are obscure it may be that the phenomenon has its origin in a more general requirement to narrow or simplify neural pathways, allowing the individual to make decisions more rapidly. Speed of decision-making can, at least in theory, improve efficiency and decrease danger, allowing a relatively polyphagous animal to attain the benefits of genetically determined narrow host range (see Chapter 8).

6.7 Compulsive requirement for novelty

In at least one species, an absolute requirement for novel flavors seems to exist. The highly polyphagous grasshopper, *Taeniopoda eques*, refuses to eat any one food plant in the laboratory over a long period, and cannot be reared without having a mixture of plants available. In the field, it switches frequently between food items. Laboratory tests showed that on every one of ten foods tested with isolated individuals there was a decline in acceptability over time. Whatever the initial level of acceptability, the food was rejected by about the fifth encounter. Further study was undertaken with good quality artificial diets that were all identical but had different added harmless flavors. After two meals on food with one flavor, individuals were more likely to eat large meals on food with a novel flavor and small meals on the food with the flavor already experienced (Fig. 6.8).

In further work, it was found that after experience with one of two different odors in the absence of food, *T.eques* ate sooner and for longer on an artificial diet laced with the novel odor, than on the diet laced with the one already experienced. This may be an exaggerated example of the rather well known general phenomenon of an increased state of arousal when environmental stimuli change. It is perhaps also related to the process of sensitization. It is of course unlikely to occur in any but polyphagous species, and, of those, only species that are mobile and individually polyphagous. It is the precise opposite of an induction of preference.

6.8 Conclusions

Experience has a variety of effects on food-selection behavior. Several different neural phenomena are involved, and in any one behavioral change there may be

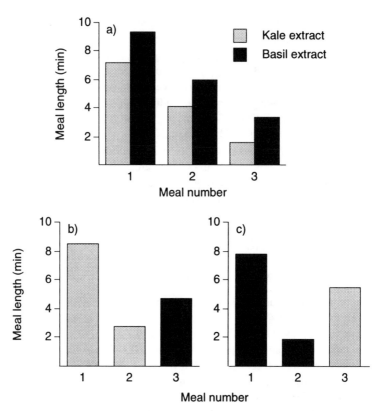

Figure 6.8. Effects of experience on feeding by the grasshopper, *Taenio-poda eques*. Individuals were given artificial diets that were identical and of good quality. To these were added chloroform extracts of either kale or basil to give unique flavors. **a)** Three successive meals by individuals on diet with either kale or basil extract. **b)** Individuals having two successive meals on diet with kale extract, followed by a meal on diet with basil extract. **c)** Individuals having two successive meals on diet with basil extract followed by a meal on diet with kale extract. The patterns demonstrate that successive meals on food with the same flavor decline in length, while a new flavor stimulates a longer meal. Novelty alone appears to be important in the feeding behavior of this species (after Bernays et al., 1992).

a mixture of peripheral change and changes in the central nervous system, and a variety of different time courses (Table 6.4). An apparent difference from vertebrates is the plasticity of the chemoreceptors, which may then govern some of the changes found, although probably not in the case of associative learning.

There has been some speculation as to the effects of behavioral plasticity on components of fitness. In one sense, it is self-evident that learning would be beneficial, yet experiments are lacking that actually measure it. There is variation

Table 6.4.   Time course for persistence of the various phenomena associated with changes in food selection behavior resulting from experience.

Chemosensory adaptation	seconds - minutes
Sensitization	minutes - hours
Nutrient-related sensillum changes	minutes - hours
Sensillum changes from plant secondary compounds	days
Habituation	hours - days
Aversion learning - nutrients	hours
Aversion learning - toxins	days
Positive associative learning	minutes - days
Induction to foods	hours to days

in learning ability, however, and presumably it is possible to select for improvements. Some arguments suggest that learning is important for dealing effectively with a variable environment. On the other hand some suggest that learning would have most value to an individual if it has a predictable environment. It is possible to resolve this apparent conflict by considering long- versus short-term predictability of the resource (in this case, food). If a reliable cue is always available in association with a particular important resource, then one might expect the evolution of fixed responses since these may be unerring and fast. If the reliability is only short-term, for example, within the life of an individual but not between generations, then fixed responses would be a disadvantage, and learning would be most useful. If the important resources are always unpredictable, or without reliable cues, then experience should probably be generally ignored. Thus, the expectation is that learning would be most favored when within-generation predictability of the resource is high and between-generation predictability is low. Table 6.5 summarizes these ideas.

Recently, there has been some exploration of the possibility that learning may, in itself, influence the direction of evolutionary change. An example may be given in the case of a female butterfly which is encountering plants other than

Table 6.5.   The selective response to be expected from insects habitually experiencing different degrees of predictability in their food/oviposition resource. The table shows the results of interactions between and within generations, for a particular species of animal.

		Within-generation predictability	
		LOW	HIGH
Between-generation predictability	LOW	Ignore experience	learn
	HIGH	Ignore experience	evolve fixed response

its normal host plants for oviposition. A large egg load may result in her decision to oviposit on nonhosts (see Chapter 5). If the act of oviposition is an unconditioned stimulus for learning, she may then continue to oviposit on the same plant. Such persistent oviposition on a novel or nonpreferred host, may increase the likelihood of some offspring accepting the new plant and growing well on it. It could be the first stage of a host shift. The phenomenon involves a physiological state variable that alters acceptance levels of a plant, combined with learning, and a resultant increase in selection for larvae with improved ability to utilize a novel host.

Further reading

Bernays, E.A. and Bright, K.L. 1992. Dietary mixing in grasshoppers: a review. Comp. Biochem. Physiol. A 104:125–131.

Blaney, W.M., Schoonhoven, L.M. and Simmonds, M.S.J. 1986. Sensitivity variations in insect chemoreceptors; a review. Experientia Exp.Appl. 42: 13–19.

Jaenike, J. and Papaj, D. 1992. Behavioral plasticity and patterns of host use by insects. *In* Roitberg, B.D. and Isman, M.B. (eds.) *Insect Chemical Ecology. An Evolutionary Approach*. Chapman & Hall, New York, pp. 245–264.

Krasne, F.B. 1984. Physiological analysis of learning in invertebrates. *In* Reinoso-Suarez, F. and Ajmone-Marson, C. (eds.) *Cortical Integration*. Raven Press, New York, pp. 53–76.

Papaj, D.R. and Lewis, A.C. (eds.) 1992. *Insect Learning*. Chapman & Hall, New York.

Papaj, D.R. and Prokopy, R.J. 1989. Ecological and evolutionary aspects of learning in phytophagous insects. A.Rev.Entomol. 34: 315–350.

Stephens, D.W. 1991. Change, regularity and value in the evolution of animal learning. Behav.Ecol. 2: 77–89.

Stephens, D.W. and Krebs, J.R. 1986. *Foraging Theory*. Princeton University Press, Princeton.

Szentesi, A. and Jermy, T. 1989. The role of experience in host plant choice by phytophagous insects. *In* Bernays, E.A. (ed.) *Insect-Plant Interactions*, vol. 2. CRC Press, Boca Raton, pp. 39–74.

References (* indicates review)

Habituation

*Jermy, T., Bernays, E.A. and Szentesi, A. 1982. The effect of repeated exposure to feeding deterrents on their acceptability to phytophagous insects. *In* Visser, J.H. and Minks, A.K. (eds.) *Insect-Plant Relationships*. Pudoc, Wageningen, pp. 25–30.

Jermy, T., Horvath, J. and Szentesi, A. 1987. The role of habituation in food selection of lepidopterous larvae: the example of *Mamestra brassicae*. *In* Labeyrie,V., Fabres, G. and Lachaise, D. (eds.) *Insects-Plants*. Junk, Dordrecht, pp. 231–236.

Szentesi, A. and Bernays, E.A. 1984. A study of behavioural habituation to a feeding deterrent in nymphs of *Schistocerca gregaria*. Physiol.Entomol. 9: 329–340.

Usher, B.F., Bernays, E.A. and Barbehenn, R.V. 1988. Antifeedant tests with larvae of *Pseudaletia unipuncta*: variability of behavioral response. Entomologia Exp.Appl. 48: 203–212.

Sensitization

Barton Browne, L., Moorhouse, J.E.and Gerwen, A.C.M. van 1976. An excitatory state generated during feeding in the locust, *Chortoicetes terminifera*. J.Insect Physiol. 21: 1731–1735.

Bernays, E.A. and Chapman, R.F. 1974. The regulation of food intake by acridids. *In* Barton Browne, L. (ed.) *Experimental Analysis of Insect Behaviour*. Springer Verlag, Berlin, pp. 48–59.

Dethier, V.G., Solomon, R.L. and Turner, L.H. 1965. Sensory input and central excitation and inhibition in the blowfly. J.Comp.Physiol.Psychol. 60: 303–313.

Associative Learning or Conditioning

Bernays, E.A. and Raubenheimer, D. 1991. Dietary mixing in grasshoppers: changes in acceptability of different plant secondary compounds associated with low levels of dietary protein. J.Insect Behav. 4: 545–556.

Bernays, E.A. and Wrubel, R.P. 1985. Learning by grasshoppers: association of colour/light intensity with food. Physiol.Entomol. 10: 359–369.

*Blaney, W.M. and Simmonds, M.S.J. 1987. Experience: a modifier of neural and behavioural sensitivity. *In* Labeyrie, V., Fabres, G. and Lachaise, D. (eds.) *Insects-Plants*. Junk, Dordrecht, pp. 237–241.

Jaenike, J. and Papaj, D. 1992. Behavioral plasticity and patterns of host use by insects. *In* Roitberg, B.D. and Isman, M.B. (eds.) *Insect Chemical Ecology. An Evolutionary Approach*. Chapman & Hall, New York, pp. 245–264.

Papaj, D.R. 1986. Conditioning of leaf shape discrimination by chemical cues in the butterfly, *Battus philenor*. Anim.Behav. 34: 1281–1288.

Papaj, D.R. and Prokopy, R.J.1986. Phytochemical basis of learning in *Rhagoletis pomonella* and other herbivorous insects. J.Chem.Ecol. 12: 1125–1143.

Simpson, S.J. and White, P.R. 1990. Associative learning and locust feeding: evidence for a "learned hunger" for protein. Anim.Behav. 40: 506–513.

Stanton, M.L. 1984. Short-term learning and the searching accuracy of egg-laying butterflies. Anim.Behav. 32: 33–40.

Traynier, R.M.M. 1986. Visual learning in assays of sinigrin solution as an oviposition releaser for the cabbage butterfly, *Pieris rapae*. Entomologia Exp.Appl. 40: 25–33.

Food aversion learning

*Bernays, E.A. 1992. Food aversion learning. *In* Lewis, A.C. and Papaj, D. (eds.) *Insect Learning*. Chapman & Hall, New York, pp. 1–17.

Bernays, E.A. and Lee, J.C. 1988. Food aversion learning in the polyphagous grasshopper *Schistocerca americana*. Physiol. Entomol. 13: 131–137.

Bernays, E.A. and Raubenheimer, D. 1991. Dietary mixing in grasshoppers: changes in acceptability of different plant secondary compounds associated with low levels of dietary protein. J.Insect Behav. 4: 545–556.

Blaney, W.M. and Simmonds, M.S.J. 1985. Food selection by locusts: the role of learning in rejection behaviour. Entomologia Exp.Appl. 39: 273–278.

Bright, K.L. and Bernays, E.A. Distinctive flavors influence mixing of nutritionally identical food by grasshoppers. Chem.Senses 16: 329–336.

Champagne, D.E. and Bernays, E.A. 1991. Phytosterol unsuitability as a factor mediating food aversion learning in the grasshopper *Schistocerca americana*. Physiol. Entomol. 16: 391–400.

Dethier, V.G. 1980. Food aversion learning in two polyphagous caterpillars, *Diacrisia virginica* and *Estigmene congrua*. Physiol.Entomol. 5: 321–325.

Lee, J.C. and Bernays, E.A. 1990. Food tastes and toxic effects: associative learning by the polyphagous grasshopper *Schistocerca americana*. Anim. Behav. 39: 163–173.

Induction of preference

Blaney, W.M and Simmonds, M.S.J. 1987. Experience, a modifier of neural and behavioural sensitivity. *In* Labeyrie, V., Fabres, G. and Lachaise, D. (eds.) *Insects Plants*. Junk, Dordrecht, pp. 237–241.

De Boer, G. and Hanson, F. 1984. Food plant selection and induction of feeding preferences among host and non-host plants in larvae of the tobacco hornworm *Manduca sexta*. Entomologia Exp.Appl. 35: 177–194.

Hanson, F.E. 1976. Comparative studies on induction of food choice preferences in lepidopterous larvae. Symp.Biol.Hung. 16: 71–77.

Jaenike, J. 1983. Induction of host preference in *Drosophila melanogaster*. Oecologia 58: 320–325.

Jaenike, J. 1988. Effects of early adult experience on host selection in insects: some experimental and theoretical results. J.Insect Behav. 1: 3–15.

*Jermy, T. 1987. The role of experience in the host selection of phytophagous insects. *In* Chapman, R.F., Bernays, E.A. and Stoffolano, J.G. (eds.) *Perspectives in Chemoreception and Behavior*. Springer-Verlag, New York, pp. 143–157.

Papaj, D.R. and Prokopy, R.J. 1988. The effect of prior adult experience on components of habitat preference in the apple maggot fly (*Rhagoletis pomonella*). Oecologia 76: 538–543.

Saxena, K.N. and Schoonhoven, L.M. 1982. Induction of orientational and feeding preferences in *Manduca sexta* larvae for different food sources. Entomologia Exp.Appl. 32: 172–180.

Compulsive requirement for novelty

Bernays, E.A. 1992. Dietary mixing in generalist grasshoppers. *In* Menken, S.B.J., Visser, J.H. and Harrewijn, P. (eds) *Proc.8th Symp.Insect-Plant Interactions*. Kluwer Academic Publications, Dordrecht, pp. 146–148.

Bernays, E.A., Bright, K.L., Howard, J.J. and Champagne, D. 1992. Variety is the spice of life: compulsive switching between foods in the generalist grasshopper *Taeniopoda eques*. Anim.Behav. 44: 721–731.

7

Genetic Variation in Host Selection

Previous chapters have emphasized the variations in feeding behavior that result from ecological interactions, developmental or physiological state, and various types of experience. By contrast, in this chapter, we present the rather limited evidence that intraspecific variation in host-plant choices can also be due to heritable differences. Additional studies have demonstrated genetic variability in growth and survival on different plants, but unless there is a behavioral component these are not treated in any detail here. Studies of heritable differences in behavior among phytophagous insects received little attention until recently, and we hope that by emphasizing the need, the genetics of host-plant selection will be examined further.

In a classic paper published in 1957, W.G.Wellington described persistent behavioral differences, including feeding activities, among individuals of the tent caterpillar, *Malacosoma pluviale*. There is no way of knowing how much of the variation was genetic but it drew attention to the great variability that can occur within insect species. Not long after, when electrophoretic techniques were used to study populations, geneticists were surprised to find that there was no single "wild type" but instead, considerable variation and heterozygosity. This led the way to the more recent studies of genetic variation and microevolution, including that of host use by phytophagous insects.

One of the major reasons for interest in genetic variation in host choice is that variability at the individual level within any population, gives a measure of the potential for selection and thus for rapid evolution. Also the study of how different populations vary and how a single population may divide into separate ones provides information on how new species arise with different host use characteristics.

As noted in earlier chapters on behavior, variation may occur at different points in the behavioral sequences that influence host selection. Foraging activity, attraction to hosts, acceptance or rejection, and relative acceptability; all may vary. There is scattered evidence that within species, genetic differences are involved at each of these different levels. In a few insect species, information on variation in chemoreceptor mechanisms provides some insight into the possible bases for the behavioral differences.

7.1 Within-population differences

Behavioral diversity within a population of insects is usual, and variation in respect of host acceptance is simply a special case of the general finding that genetic variation in almost all traits in all organisms is the norm. Before about 1975, this was not really appreciated, but now it has become an important aspect of biological research.

Single locus traits of various kinds, with associated markers, have been studied for 50 years or so. The recent addition of molecular markers has greatly enhanced our appreciation of variation present within populations. However, the more recent studies of quantitative traits, has improved the study of genetic factors more directly associated with behavior and life history. This is because behavioral traits are usually quantitative, and determined by genes at more than one locus. These studies have further demonstrated that behaviors vary as much or more than any other biological features. The interest in within-population genetic differences is partly an interest in how much variation is available for selection mechanisms to act upon, which in turn can determine the potential for rapid evolutionary change. In addition, there is considerable interest in how evolutionary changes can impact herbivory on agriculturally important plants, and how rapidly insects may adapt to utilizing resistant species or varieties.

7.1.1 Natural populations of herbivores

Increasingly, it is being found that within a population, there are groups of individuals that utilize different hosts, and it is assumed that these are genetic variants within the species. In some cases, there is evidence that the phenotypic differences really are genetically based. This has been demonstrated in a number of studies of the butterfly, *Euphydryas editha*. For example, at a study site in Nevada some females oviposit on *Collinsia* (Scrophulariaceae) while others use *Plantago* (Plantaginaceae). Heritability of this postalighting preference was demonstrated. In a similar study at a site in the Sierra Nevada in California, the same species showed three different patterns of preference within the same population. One group of insects preferred *Collinsia*, a second group preferred *Plantago*, while the third group chose both equally. In a third study, Ng and Singer found that in one site in the Sierra Nevada, there were two types of *Euphydryas editha*: "specialists" that consistently picked only some individuals of the host plant, *Pedicularis semibarbata*; and "generalists" that accepted all the known host plants. Rearing studies indicated that the differences in oviposition behavior were heritable.

Another well documented example is that of *Papilio zelicaon*, a butterfly widespread in western North America. Populations within a small part of its range in Oregon have preferences for different hosts, various species of Apiaceae (=Umbelliferae) and citrus (Rutaceae). Females were collected from a single site and tested for their acceptance of different host species. Their progeny were tested as adults with a choice of the same hosts. They showed preferences similar to those exhibited by their mothers, although sometimes eggs were also laid on the alternative hosts.

The foliage-feeding larvae of the moth, *Alsophila pometaria*, utilize deciduous angiosperms in numerous families. In Long Island, New York, some parthenogenetic populations are associated with stands of red maple and others with oak. Egg masses are laid in winter and young larvae hatch in spring and feed on the newly developing foliage. In experiments where egg masses from these two genotypes were placed on the same or the alternative foliage, hatchlings were examined for their propensity to disperse. This assay is a measure of the degree to which such larvae reject or accept the foliage. Dispersal from maple foliage was greater for the oak-associated genotype than the maple-associated genotype, although the response to oak did not differ between genotypes (Fig. 7.1). In this incipient divergence in host use, genetic differences in behavior were found but there were no differences in digestive adaptation.

Within populations where genetic polymorphism in host use occurs, ecological factors such as changes in plant phenology may reduce interbreeding between genotypes. Low mobility or assortative mating by the insects would additionally reduce the likelihood of interbreeding, and when genotypes become pronounced in their differences they are often called "races". These could be the expected precursors of new species that arise within a region, a process known as sympatric speciation.

The formation of host races, or conspecific populations specialized on alternative hosts, appears to have occurred in the apple maggot fly, *Rhagoletis pomonella*. In this species, mating occurs on or near the fruit, and if two genotypes using two different fruit types mate on their own host fruits, there will obviously be reduced gene flow between the genotypes. In *R. pomonella*, this reduced gene flow is enhanced by temporal differences in availability of the two main hosts, hawthorn and apple. Local temperatures differentially influence the timing of adult eclosion, such that adults of one genotype coincide with the fruiting phenology of one host, and adults of the second genotype coincide with the other. However, there are also differences in preference for the different hosts. Experiments have shown that there are distinct differences which are largely experiential but partly genetic. Thus among the naive insects, those of apple origin chose apple significantly more often than those of hawthorn origin (Fig. 7.2).

When it comes to the study of detailed behavioral responses and their genetic

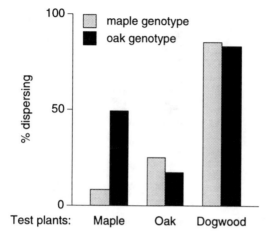

Figure 7.1. Dispersal of newly hatched *Alsophila* larvae from three different potential host plants: maple, the host of one genotype; oak, the host of the other genotype; and dogwood, a nonhost. Dispersal gives a measure of unacceptability of the plant and it can be seen that the oak genotype rejects maple much more than does the maple genotype (after Futuyma et al., 1984).

analyses, there are considerably less data. The most comprehensive study has been carried out with two *Papilio* species. The butterflies were collected from one site and experiments were carried out with individuals from isofemale lines (offspring from single females). Individual *Papilio zelicaon* and *P. oregonius* females were allowed to oviposit on four different plant species which are ranked differently by the two butterflies. Both laid the majority of eggs on their usual host-plant species: two umbels in the case of *P. zelicaon*, and a composite in the case of *P. oregonius*. However, there were differences between the isofemale lines within a species. The rank order of preference did not change but lines differed in the degree to which females laid some eggs on plants other than their usual hosts (Fig. 7.3).

Within lines there were also differences. Both lines of *P. oregonius* had a few females that distributed their eggs among the four plant species instead of laying them almost exclusively on the composite. Among *P. zelicaon*, some females laid most eggs on *Lomatium*, while others laid most eggs on *Cymopterus*. Most laid some eggs on *Foeniculum*, and a few females laid some of their eggs on the composite, *Artemisia* (Fig. 7.4). This work indicates that local distributions of insects on plant species could reflect a complex mosaic of genotypes. Alternative and novel hosts are acceptable to some of the individuals and there is potential for rapid host shifts.

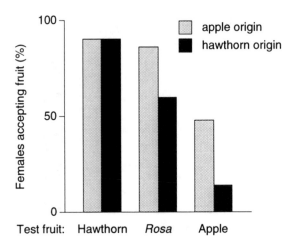

Figure 7.2. Innate preferences of *Rhagoletis pomonella* females for oviposition substrates. Puparia were collected so that flies had no experience of any fruit prior to the experiments. Individual flies were offered single fruit and either accepted it within a certain time and laid an egg, or they rejected it. The figure shows that the flies of apple origin are more likely to accept apple or another host, *Rosa rugosa*, than were the flies of hawthorn origin (after Prokopy et al., 1982).

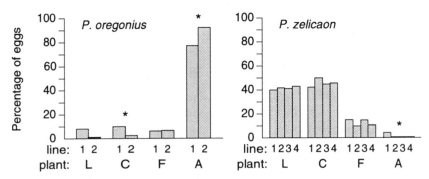

Figure 7.3. Percentages of eggs laid on four plant species by isofemale lines of *Papilio oregonius* (lines 1 and 2) and *P.zelicaon* (lines 1-4). Plants are: L, *Lomatium grayi* (Apiaceae); C, *Cymopterus terebinthinus* (Apiaceae); F, *Foeniculum vulgare* (Apiaceae); and A, *Artemisia dracunculus* (Asteraceae). The four plant species were offered to all females in choice experiments. Plants were in random arrays on all days in the experiments. The two butterfly species have different preferences but within each species the different lines also show some significant differences (indicated by *), which relate mainly to their oviposition on *A.dracunculus* (after Thompson, 1988b).

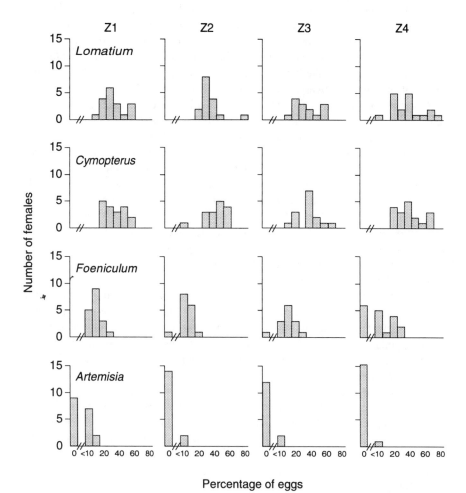

Figure 7.4. Distribution of percentages of eggs laid on four plant species by individual females within the four isofemale lines (Z1 to Z4) of *Papilio zelicaon* shown in Fig. 7.3. Butterflies had a choice of the four plants and laid eggs on one, two, three or on all four. Individual butterflies from all four lines usually laid most of their eggs on the two normal host plants, *Lomatium* and *Cymopterus*. None of the individuals laid a majority of eggs on the novel *Foeniculum*, and some females in Z4 avoided this plant. Females of the Z2, Z3 and Z4 lines mostly laid few or no eggs on the novel plant *Artemisia*, but a large number of Z1 females laid some eggs on this plant. That is, some individuals have a relatively higher acceptance level of the unusual hosts (after Thompson, 1988b).

In additional experiments, F_2 individuals from crosses of isofemale lines within each of the species and from crosses of the two *Papilio* species were obtained and tested. The results indicated that oviposition preference was affected by one or more loci on the X chromosome with additional effects from one or more loci not on the X chromosome. Since, in Lepidoptera, the male is XX and female XO, hybrid females inherit the X chromosome only from the father, so that paternal inheritance governs the oviposition behavior of the female.

Considerable variation is characteristic of all aspects of feeding behavior in phytophagous insects and the contribution of genetic differences to these details is generally unknown. For example, in one study on the Colorado potato beetle, *Leptinotarsa decemlineata*, individual insects within populations showed distinct differences in the details of their behavior. In a Canadian population, most of which had a low preference for *Solanum elaeagnifolium*, there were some individuals that consistently and readily ingested the food. They had long meals, short examination times on the leaf, short decision times after biting, and ate at one site rather than at several during the course of a meal. This indicated that they perceived the plant as acceptable prior to feeding. The beetles were naive adults in this study and the differences between individuals were assumed to be innate although there were no genetic studies.

In a quantitative genetic study of the polyphagous grasshopper, *Melanoplus differentialis*, it has been shown that in some families of the insect, changing the food plants is acceptable, while in others the individuals prefer to feed on single plants. Further, there were family differences in the way in which food plants were mixed: some took mixtures of plants within meals, and others obtained mixtures by eating different plants in different meals.

Studies with *Drosophila* have given new insights into what must be studied further in phytophagous species. Genetic variation for settling responses on different types of food and, independently for oviposition, exists in natural populations of *Drosophila tripunctata*. In a choice of tomato and mushroom in the field, one strain was preferentially attracted to tomato and the other to mushroom (Fig. 7.5a). The F2 individuals derived from inbred lines of these were intermediate in their responses. On the other hand, females of the strain which land preferentially on tomatoes mostly preferred to oviposit on mushroom, while those attracted to mushroom showed no preference between the two substrates (Fig. 7.5b). Furthermore, when flies of the two parental strains are classified by bait type at which they were captured, it is clear that the females captured at tomatoes in the field, showed a stronger preference for mushrooms in the oviposition test than did those captured at mushrooms. From this it is apparent that oviposition behavior of the flies is independent of their initial substrate choice in the field, and the two behaviors are under independent genetic control. The oviposition behaviors of the reciprocal crosses were similar, indicating that the contribution of the X chromosome to variation in behavior is insignificant, and the two loci are not linked. The genetic complexity in this *Drosophila* case

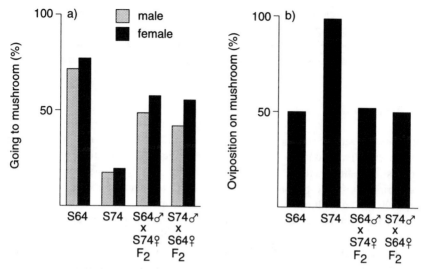

Figure 7.5. Differences in attraction and oviposition behavior of two strains of *Drosophila tripunctata*. Insects had a choice of mushrooms and tomatoes. Only responses to mushroom are shown. **a)** Percent of females and males attracted to mushrooms. Strain S64 prefers to go to mushrooms; only about 20% of S74 go to mushrooms (they prefer tomatoes). Among progeny of both crosses about 50% go to mushroom (they show no preference for either mushroom or tomato). **b)** Percent of females laying eggs on mushrooms. S74 lays nearly all its eggs on mushrooms even though it is more strongly attracted to tomato (a); S64 and progeny of the two crosses only lay about 50% of their eggs on mushroom even though S64 is strongly attracted to mushroom (a) (after Jaenike, 1986).

should lead to caution in using a single phase of host search as the indication of final oviposition choice.

With foliage-feeding insects no similar studies have been carried out to determine if attraction (for example, by visual or odor cues), is under separate genetic control from host acceptance upon contact. Considering the different sensory modalities being used however (see Chapter 4), it might be expected that this would be the case.

In conclusion, it would appear from the growing number of studies showing genetic differentiation in host use within populations, that it may be universal. This would, in turn, suggest that sympatric speciation, associated with differential use of hosts, may have been quite common, and perhaps one of the elements leading to the great number of species of plant-feeding insects.

One of the interesting questions that arises out of the existence of variation in respect of host selection behavior and host use generally, is why this feature should be so variable. One possible contributing factor in natural systems is that the plants are not homogeneous themselves. The variation in the insects may reflect changing selection pressures due to continually changing host populations. Another contributing factor could be changes in availability of different plant taxa due to geographical range changes in either insects or plant hosts. The insect may be undergoing constant selection to keep up and thus may rarely come into equilibrium.

7.1.2 Biotypes

Biotypes are strains of insects with inherited differences in their ability to use host species. Traditionally, they have been identified by differential use of particular host varieties, without any genetic analysis. As the genetic mechanisms conferring resistance become better known, the term may become less useful. Typically, in agriculture, whenever a resistant crop variety is no longer resistant, a new pest "biotype" is said to have arisen. The greenbug on cereals, *Schizaphis graminum*, for example, has eight biotypes, while the Hessian fly on wheat, *Mayetiola destructor*, has 12 biotypes. In the case of the greenbug, serious problems arose with this pest in the 1930s and various wheat varieties were tested for resistance during the 1940s. By 1945 differences among greenbugs in their ability to damage different wheat varieties had been noted, though they were not labelled biotypes. In 1955, a wheat variety was developed that was resistant to the greenbug, but by 1961 insects had adapted to this variety and these virulent genotypes quickly became a major pest and were labelled "biotype B" to differentiate them from the original, "biotype A." In 1968, greenbugs were found damaging sorghum which had previously been a poor host, and these were labelled "biotype C." Changes continued and still continue. The situation is complex as indicated by some of the combinations of abilities of seven biotypes shown in Table 7.1.

There has been considerable controversy over the definition of biotype since it says nothing about what is different, and whether it is quantitative or a major gene change. Usually the genetic makeup of the biotypes is unknown, and, clearly, in many cases, they are easily changed from one to another by selection experiments (see below). The term is best used to signify a group of insects that have similar responses to a particular trait in certain plant varieties. The biotype may actually consist of genetically different individuals, however, and these may show differential abilities on a different crop or variety. For example a biotype unable to grow on one variety may consist of individuals with differential ability to grow on a second variety.

The evolution of biotypes that are able to use new or previously resistant crop varieties is well studied in the rice brown planthopper, *Nilaparvata lugens*. Artificial selection experiments were carried out on several different biotypes. All showed variation for feeding and performance, and changes from one to another could be brought about in 11 generations. For example, an Australian population was examined for variation in feeding rate, determined by honeydew

Table 7.1.   Resistance (+) and susceptibility (−) of plants to greenbug (Schizaphis graminum) biotypes. The table reveals complex interactions between biotypes and hosts, since few plant varieties are resistant to all biotypes (after Wilhoit, 1992).

Plants	Greenbug biotypes					
	B	C	E	F	G	H
Wheat varieties						
TR64	+	+	+	+	+	+
C19058	+	+	+	−	+	+
Amigo	−	−	+	+	+	+
Largo	+	−	−	+	+	−
10563B	−	−	−	+	+	−
Sorghum varieties						
Wheatland		+	+	+	−	−
KS30	−	−	+	−	−	−
P8493		−	−	−	−	−
Piper	−	+	+	−	−	−
Oats varieties						
Nora	+	+	+	+	+	+
C11580	+	−	−	+	+	+

production rates, on several different rice varieties. The differences among individuals were large. Some fed for quite long times even on the resistant varieties (Fig. 7.6). Survivorship on the resistant TN1 increased by a factor of four in 11 generations (Fig. 7.7), even though there was relatively little variation initially.

Trying to evaluate exactly what might be the genetic basis for this resistance is almost impossible. Further, even knowing that we have a relatively stable situation, consideration has to be given to how best to deploy the use of the new varieties. Modelling promises to allow us to deal with very complex situations that have many parameters, and insect populations that have different genetic makeups. F. Gould has employed this approach with promising results, to both sexually reproducing pests and to asexually reproducing aphids. From these models he could predict the duration of effective resistance in different regimes of planting, and with different genetic parameters of the pest insects. Thus working back from such models to what is actually possible in field studies, and incorporating more ecological data, the relative importance of different factors can be established.

7.1.3 Genetic variation demonstrated by selection experiments

Laboratory experiments provide some further information on the available genetic variation upon which selection can act, and the possibility that host-use patterns may be altered under strong selection pressure. One of the first such experiments was with the phytophagous mite *Tetranychus urticae* by F. Gould. These were collected in one place and reared on bean plants for 16 generations. They were

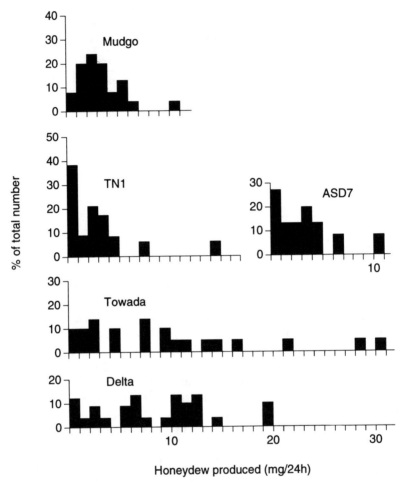

Figure 7.6. Feeding, measured as honeydew production, by individuals of an Australian population of *Nilaparvata lugens* tested on five different rice varieties. Mudgo, TN1 and ASD7 were fairly resistant (relatively little honeydew was produced), compared with Towada and Delta (after Claridge and den Hollander, 1982).

then divided into groups. Some were kept on beans alone, and some on a mixture of cucumber and beans. In the latter groups the mites fed first on beans, which they preferred, and became very numerous. After consumption of the beans, they moved onto cucumber and the survivors were used to start a new colony. After some months, the line selected on cucumber became significantly better able to grow on this marginal host as well as other marginal hosts, and slightly less able to use beans. The new, selected line could also be reselected on beans only, whereupon they eventually lost their abilities to use cucumber.

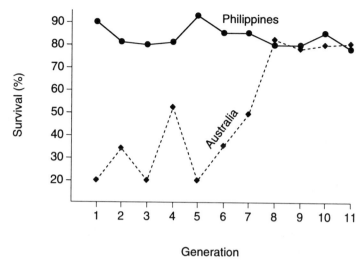

Figure 7.7. Percentage survival of the rice brown planthopper, *Nilaparvata lugens*, from Australia and the Philippines (biotype 1), each selected on TN1 for 11 generations. Biotype 1 maintained its ability to use TN1, while the Australian population increased its ability to use this variety over only eight generations (after Claridge and den Hollander, 1982).

One of the interesting factors demonstrated by this experiment is that the genetic variability for change was apparently present in the population collected in one field. The results also indicate that host-range evolution can be a rapid process, and involve not just adaptation to the new plant, but also cross-adaptation to additional plants. However, little is known about the role of behavioral preferences versus digestive physiology and/or toxicology in the changes that occurred.

In an elegant experiment with the bruchid beetle *Callosobruchus maculatus*, groups of individuals from one source were selected for larval adaptation to, and female oviposition preference for, different species of hosts. The experiments lasted over eleven generations. In one example of the experiments, adults were offered both azuki beans, normally preferred for oviposition, and pigeon peas, less preferred. The azuki beans were discarded after oviposition, so that females which had laid eggs on them had no progeny. This provided a strong artificial selection for choice of pigeon peas for oviposition by the females (Fig. 7.8). In another experiment, beetles were given one or the other food for oviposition and development, without any choice. Survivorship and growth of larvae were tested on a subset of each host in each generation. Growth and survivorship of larvae did not change. These two experiments with *Callosobruchus* showed that genetic change occurred in the behavioral response to these hosts but not in the adaptedness of the larvae to deal with the food. This may suggest that the diets of phytophagous insects could be evolutionarily more labile at the sensory level than at the postingestive level.

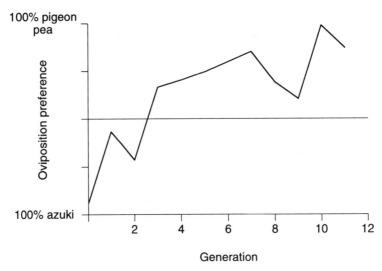

Figure 7.8. Changing oviposition preference for pigeon pea by the bean weevil *Callosobruchus*, over 11 generations of selection. Adult females were given a choice between azuki bean and pigeon pea in each generation. Any eggs laid in azuki bean were killed. The original preference for azuki bean was changed to a preference for pigeon pea (after Wasserman and Futuyma, 1981).

The insect neurophysiology and/or plant features underlying the changes in preference are unknown, although it is usually assumed that the differences are related to presence or absence of certain secondary metabolites. There is one interesting example, however, of selection for preferences in relation to protein or carbohydrate. This was carried out on *Drosophila melanogaster*, and although this species is not strictly phytophagous, the possibilities for selection make it particularly interesting. Larval insects were selected over 44 generations for choice of high protein content food versus high energy content food. In the first case, larvae that moved to the high energy food were killed, while in the second case, larvae that moved to the high protein food were killed. In this way, lines were selected that preferred high protein over high energy food and lines were selected preferring high energy over high protein. There were significant responses to selection in both directions. In other words, both lines showed a significantly higher preference for the selected food type than did the control line. Under the laboratory conditions, however, fitness-related characters were always better on the high protein food. Thus selection for behavioral changes in relation to food choice may occur, even if the selection pressure is something other than the nutritional value of the food. In translating this to a field situation, one could envisage a mortality factor selecting for choice of a host species that is not nutritionally optimal. Factors such as predation or parasitism may be of greater significance than an optimal diet, and clearly selection for behaviors that involve choosing suboptimal diets is possible.

In conclusion, the experimental data demonstrate the importance of selection. It can and does act on food choice behavior by insect herbivores, which provides some evidence that the observed variation in nature has adaptive significance, as opposed to being just neutral noise.

7.2 Between-population differences

We have seen that individuals within a population of any species show genetic variation for host use. Further, there may be factors that tend to isolate different host-use types from one another so that they develop into races or biotypes with reduced gene flow between them. These, in turn may become sufficiently different as to evolve into different species. There is, however, a continuum between the variation seen within a species in a specified region, and the variation seen between populations of a species occupying different regions. Nonetheless, there is a greater likelihood of consistent genetic differentiation between populations of a species in different regions, than within a region.

Species populations covering a large area may be genetically continuous or they may be subdivided into subunits or demes. In the first case, gene frequency changes due to local selection are likely to be overwhelmed by movement between regions. In other words, gene flow prevents the buildup of local differences. Genetic differences between populations can readily arise if gene flow among them is prevented. If there are small demes, genetic change may be relatively fast. This is because genetic drift, together with inbreeding, can ensure that new or rare alleles are maintained at higher levels than would be the case if they were repeatedly masked by influxes of individuals from more distant parts of the range. Selection then has the opportunity to act on locally common alleles that are recessive elsewhere.

Opportunity for the evolution of genetic differences between populations of plant-feeding insects varies with such factors as population size and geographic range, migratory capability and host-plant range. The rapidity of change depends also on the generation time. It has been calculated that a mutation producing a dominant allele with a 10% increase in fitness, would reach a frequency of 90% in the population in only 100 generations. For a recessive allele, the change in frequency would take about ten times as many generations. For a continuously breeding insect species of small size, the time frame for a major change in one population (assuming the 10% fitness change) might be just a few years. The potential for change of host plant, to make use of new or different resources is therefore considerable, and a study of the current genetic variation can give us indications of what selection can act upon now.

This section deals with the relatively few situations that have been examined in depth, from species with continuous populations to those in small demes.

7.2.1 Restriction of host use due to availability

Many species of phytophagous insects show population differences in host-use patterns. In most cases, these population differences appear to reflect abundance, reliability or quality of the plants in different areas, and genetic components of these differences are unknown. Among polyphagous species, this is often simply a case of using the available plant species in particular regions. For example, the agricultural pest, *Helicoverpa zea*, is known by different common names according to the crops it damages. It is called the tomato fruitworm in tomato-growing areas and the corn earworm in the corn belt of the USA. *H. zea* is very mobile and genetic differences between these populations have not been identified.

It is well known that the similar, highly polyphagous noctuid moth, *Heliothis virescens*, uses different major hosts in different regions of the United States. For example, it is a pest of tobacco in the southeast, cotton in the south, and vegetables in California. The most common crops are those most used as hosts. Like *H. zea*, it is highly mobile, although the flight range of individual moths is unknown. A study of oviposition in moths from various regions was unable to identify genetic differences between the mainland populations, but there was a difference between moths on mainland America and moths in the Virgin Islands. This demonstrates the fact that if there is sufficient isolation to prevent gene flow between populations, genetic changes may follow, giving insects in the different populations innate differences in the acceptability of certain plants.

Species of insects occurring over a wide geographic range are not necessarily polyphagous, but often must use different hosts. Several species of butterflies use different hosts in different regions and the separate populations are likely to be subjected to different selection pressures. If populations become geographically isolated, and genetic differences increase, new species may arise. This is generally considered to be the usual way in which speciation occurs and is known as allopatric speciation. Indeed, many early examples of intra-specific population differences among insects cited in the literature have since been considered to consist of separate species.

The genus *Papilio* has been the subject of study by several recent authors, and there are different degrees of genetic separation between populations in the species groups examined. A detailed study of *P. zelicaon* showed that this western United States species has populations that differ in the host plants on which the females oviposit. Host use is correlated with general availability of the plants, and, in one study, females were observed to select *Lomatium grayi* and *Cymopterus terebinthinus* (both in Apiaceae) in one area, *Angelica lucida* and *Foeniculum vulgare* (also in Apiaceae) in another, and *Citrus* spp. (Rutaceae) in a third. Furthermore, the differences were maintained when females were collected from the field and tested in the laboratory, and the differences were found to be genetically based.

Papilio cresphontes, the giant swallowtail, is basically a specialist on plants in the family Rutaceae, but populations utilize different food-plant species in various parts of its range. *Citrus* species are favored in Florida, where citrus is widely grown. In the northern United States, where there is no citrus, the prickly ash *Zanthoxylum americanum* is the primary host plant. Experiments showed that there are basic physiological differences in the populations and, although behavior was not separately examined, it was assumed that oviposition preferences are involved.

In the *Papilio glaucus* species complex, differentiation has gone further and different subspecies can be identified. *P. glaucus glaucus* in southeastern USA has a relatively wide range of host trees, while *P. glaucus canadensis* in northern USA and Canada utilize different, and fewer host species. There are clear differences in host selection behavior, when individuals are given choices, although at the zone of overlap both prefer black cherry, *Prunus serotina*. Hybrids between the two accept hosts of both parents (Fig. 7.9).

Population differences may be quite local, depending on the size and mobility of the insect concerned. The black pine leaf scale insect, *Nucolaspis californica*, is an extremely sedentary herbivore recorded from about ten pine and fir species. As with other coccids, the dispersal phase is the newly hatched nymph or crawler,

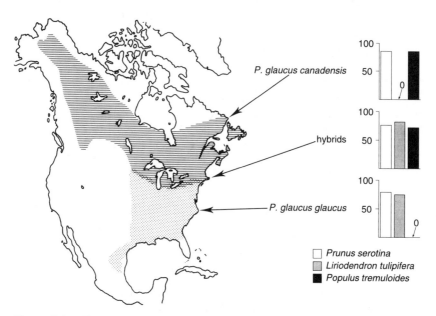

Figure 7.9. The geographic range and percent larval survival to second instar on three key food plants of the two subspecies of *Papilio glaucus*. Hybrids accept all three food equally readily although the parental populations show different preferences (after Scriber, 1982).

which is carried to new hosts by wind. It is common for a few crawlers to become established on a nearby tree and the population increases gradually over many generations. Over time, the ability to establish colonies on different hosts declines, suggesting that the deme has become specialized on its own host. Also, there is little gene flow between colonies on different trees, relative to the selection pressure. There is probably intense selection on the insects on individual trees so that demes develop with genetic adaptations that are appropriate for the chemistry of each individual tree. Only those demes with adequate genetic variation will be able to produce crawlers able to survive on a different tree. Although it is assumed that digestive adaptation is occurring, it is likely also that acceptability of the host tree and sensory/neural tolerance of its chemistry must play a part, especially since much of the within-tree mortality is at the crawler stage.

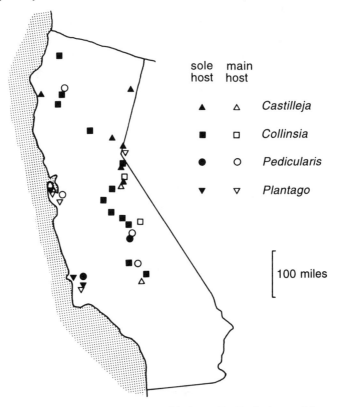

Figure 7.10.   Host-use patterns of the butterfly, *Euphydryas editha*, in different parts of California. Symbols indicate the sole or main host used in each area studied. Although this species uses only plants containing iridoid glycosides, different plant species are used by different populations. The host used in the field is usually the most abundant, the best quality or in the best microhabitat (after Singer, 1993).

Few studies have concentrated on the genetics of behavior. M. Singer was one of the first to document highly significant differences in oviposition preference between different populations of herbivores. He showed that different populations of the checker spot butterfly, *Euphydryas editha*, in California, preferred to oviposit on different species of plants (Fig. 7.10). Genetic differences are indicated, since a correlation was found between host choice of mothers and their daughters.

A detailed behavioral study has been made of the foraging behavior of two different populations of the cabbage butterfly, *Pieris rapae*, in different geographic regions. This species is native to Europe but arrived in Australia in 1939. Very marked differences now occur in behavior between the Australian and British females, especially in the numbers of eggs laid per alightment (Table 7.2, Fig. 7.11). When the two populations were kept together in the laboratory on the same food and under the same conditions, they maintained the differences in successive generations. The major genetically-based behavioral differences are unaccompanied by any morphological or developmental differences. This suggests a remarkable lability in behavioral characters, and indicates that behavioral divergence of geographically separate populations may provide the first indication of change. The adaptive explanation probably relates to differences in the weather, since in Britain, there may often be days that are too cold or too wet for flight, thus search time would be limited.

A detailed laboratory study was made of differences in feeding patterns in three different populations of the Colorado potato beetle, *Leptinotarsa decemlineata*. Insects from Alberta, Arizona and Maryland were compared on four solanaceous hosts. Acceptance of the hosts varied with populations (Table 7.3). For example, most beetles from Arizona did not feed on *Solanum tuberosum* or *Lycopersicon esculentum*, while beetles of all three populations preferred the supposed ancestral host *Solanum rostratum*. Beetles consistently sampled the leaves of less-preferred hosts longer and more frequently. They did this at two different levels; they

Table 7.2. Behavioral differences between female cabbage butterflies, Pieris rapae, *in Australia and Britain (after Jones, 1987).*

Australia	Britain
Behaviors	
Moved more to new plants	Resettled on same plant more often
Fewer alightments/min	More alightments/min
Alighted more on non-hosts	Alighted little on non-hosts
Visited smaller hosts	Visited larger hosts
Results of behaviors	
Laid eggs more slowly	Laid eggs more rapidly
Used more individual hosts	Used fewer hosts

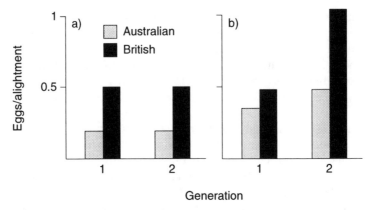

Figure 7.11. Number of eggs deposited per alightment by the cabbage butterfly, *Pieris rapae*. The Australian and British genotypes are significantly different when tested immediately after being collected from the field (generation 1), and the differences are maintained in their offspring (generation 2) (after Jones, 1987). **a)** Slow-developing females. **b)** Fast-developing females.

spent more time touching the food before biting into it (the preincision level), and they spent more time macerating the food between the mandibles before any ingestion took place (the prefeed maceration level). Furthermore, they fed at more sites on the less preferred plant species. The graded change in plant sampling behavior suggests that a complex of stimuli are involved. Although the genetic component of these differences was not specifically investigated, naive adults, which had had no contact with food were used for the tests, and the assumption is that the responses were at least innate.

Table 7.3. Feeding behavior of naive adult Leptinotarsa decemlineata *beetles from three different populations; Alberta (Al), Arizona (Az) and Maryland (Md) on four different host plants:* Solanum rostratum, Solanum tuberosum, Solanum elaeagnifolium *and* Lycopersicon esculentum *(after Harrison, 1987).*

Food Plants	Percent of beetles feeding on plant			Amounts eaten in first meal (mm^2)		
	Al	Az	Md	Al	Az	Md
S.rostratum	100	100	100	82	33	28
S.tuberosum	100	18	100	72	<1	45
S.elaeagnifolium	92	100	92	43	44	41
L.esculentum	92	9	75	5	<1	21

7.2.2 Expansion of host use due to availability

Shifts in host use by relatively restricted feeders have probably occurred many times during the evolution of insects and angiosperms, since groups of closely related insect species often differ in their use of plant species. In addition, many examples exist of phytophagous insect species that have added new hosts to their diet in the last 100 years. These are often referred to as host shifts, but they are more accurately called host-range expansions. In many cases, there is simply increased ecological opportunity, such as the recruitment of insects onto plants that have invaded a new area, or been brought into a new area by human activity. Extreme examples are among crops such as sugar cane, coffee and soybeans. With sugarcane, introduction to new regions from its origin in New Guinea, has been occurring for over 2,000 years. Recruitment of local insects has occurred wherever sugarcane was planted. Initially, polyphagous species are usually involved, but later, more specific insects tend to make use of the new plant (see Chapter 8).

Sometimes there are apparent chemical explanations for the use of novel hosts. For example the garden nasturtium, *Tropaeolum majus*, was introduced to Europe from Peru about 300 years ago, and was added to the host range of the cabbage butterflies, presumably because it contains chemicals (glucosinolates) related to those in the cabbage family. Several examples of host-range expansion have been studied among tephritid flies whose larvae feed on fruit. *Rhagoletis indifferens* in Western USA included cultivated cherries, *Prunus cerasus*, in its diet around 1913; *R. pomonella*, the apple maggot fly in eastern USA included apple, *Pyrus malus*, in its diet at a similar time. The native hosts of these species are the genera *Prunus* and *Crataegus* respectively. These two examples of host-range expansion, have occurred within the family Rosaceae, but they have led to the evolution of host races with genetically distinct characteristics in relation to host use (see below).

An interesting and well-documented example of increased host use in different populations of a specialist is found in the Colorado potato beetle, *Leptinotarsa decemlineata*. Once again, a host-range expansion has preceded some differentiation of populations with different patterns of host use. This species is indigenous to the foothills of the Rocky Mountains where it feeds on a native plant, *Solanum rostratum*. Its evolution of an ability to use potato, *S.tuberosum*, and to become adapted to grow very well on it, has enabled the beetle to expand its geographic range. It exists in most of the United States and Mexico. In North America, now, different populations use different host plants in the field: for example, *Solanum rostratum* in New Mexico and Texas, *Solanum elaeagnifolium* in Arizona, *Solanum tuberosum*, the cultivated potato, in Utah and most states north, and *Solanum rostratum* and *S. angustifolium* in Mexico. However, in all cases, the insects tested behaviorally prefer to oviposit on *S.rostratum*. Thus, in the absence of the preferred host, related species will be utilized, but the larvae vary in their ability to use them. Cross-breeding experiments between populations showed that the traits for host-adaptation are heritable. Genetic divergence was also revealed by cytogenetic analysis of chromosomal karyotypes and gel electrophoresis of isozymes. Few populations were able to survive on *S. elaeagnifolium*, but the Arizona strain was particularly insensitive to deterrent substances in this plant.

In eastern North America the plant sucking membracid, *Echenopa binota*, has races associated with different species of trees. They occur together in the same region but show marked differences in coloration and in behavior. Each preferentially oviposits in its own tree species. When given a choice of mates, females choose males from their own tree. The races are known to be genetically separate, and there is the possibility that they are in fact different species.

In conclusion, it seems that ecological opportunity leads to the inclusion of new hosts in an insect's diet. This in turn may lead to an increased geographic range and possible genetic changes across the range.

7.3 Possible mechanisms of behavioral variability

Little is known about what may be mediating the levels of acceptance or deterrence of a potential host in the cases where genetic variation has been documented. However, Wieczorek examined the chemoreceptor physiology of two genetically different strains of the noctuid caterpillar, *Mamestra brassicae*. He tested 14 secondary metabolites and in 12 of these he found significant differences in relative sensitivity between the strains as measured by the spike frequency produced by the different chemicals in chemoreceptors. Some of the data is illustrated in Fig. 7.12. It would seem that strain II is less stimulated by such deterrent compounds as salicin and strychnine, but more stimulated by some of the glucosinolates.

The sensitivity of the antennal olfactory receptor system differs between populations of the Colorado potato beetle, *Leptinotarsa decemlineata*. Beetles of the field population in the Netherlands are relatively more sensitive to *cis*-3-hexenyl acetate than those of the laboratory stock, while beetles from Utah in USA are relatively less sensitive to *trans*-2-hexen-1-ol and *trans*-2-hexenal than the other two (Fig. 7.13). The functional significance of these differences needs further elucidation but the data give some indication of the lability of the sensory system.

In the study of chemoreceptors of apparently standardized insects, variation is well known and often due to experience. Little attention has been paid to consistent differences among individuals, however, and usually the data are available only in terms of mean values. Examination of certain studies does reveal that there are differences among individuals that cannot easily be attributed to previous experience. For example, in neurophysiological studies of *Pieris brassicae* considerable differences between individuals in several parameters can be seen. Different individuals showed consistent differences in intensity of

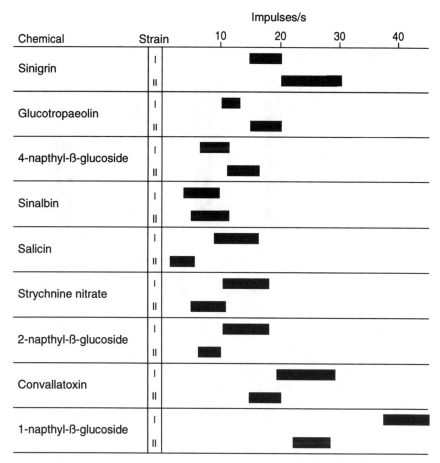

Figure 7.12. Variation in the responses of taste receptors (impulses/s) to different chemicals in two strains (I and II) of *Mamestra brassicae*. Concentration in all cases is 10^{-2} M. The bars represent confidence intervals of the medians for $P<0.05$ (after Wieczorek, 1976).

response in sensilla on the galeae to sucrose, glucose and fructose (Fig. 7.14). Furthermore, individuals showed considerable differences in the relative stimulating effectiveness of deterrents and the phagostimulant sinalbin (Fig. 7.15). In these studies, larvae tested were offspring of single females, fed on standard food, and of the same age.

Although not phytophagous, it is interesting that there are several mutants of *Drosophila melanogaster* with different types of antennal chemoreceptors missing. The individuals with missing receptors exhibit weaker electroantennograms and less marked behavioral responses to various classes of food-related volatiles compared with normal insects. Furthermore, there are many identified mutations

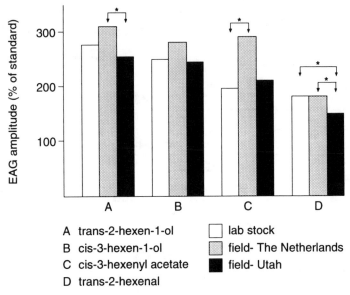

A trans-2-hexen-1-ol   ☐   lab stock
B cis-3-hexen-1-ol   ▨   field- The Netherlands
C cis-3-hexenyl acetate   ■   field- Utah
D trans-2-hexenal

Figure 7.13.   Electroantennagram (EAG) responses of Colorado potato beetles, *Leptinotarsa decemlineata*, from three different populations to four plant volatiles. Concentrations all 10^{-1} (v/v) in paraffin oil. An asterisk indicates a statistically significant difference between the columns indicated by arrows (after Visser, 1983).

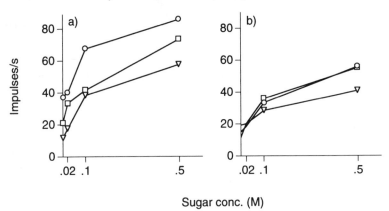

Sugar conc. (M)

Figure 7.14.   Examples of individual differences in sensory responses to two sugars by larvae of *Pieris brassicae*. Different individuals are represented by different symbols and each point is the mean of two recordings (after Ma, 1972). **a)** Response to sucrose. **b)** Response to fructose.

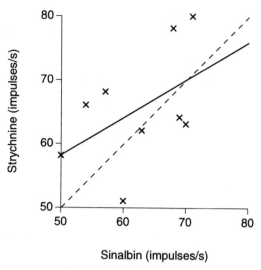

Sinalbin (impulses/s)

Figure 7.15. Firing rates (in impulses per second) of chemoreceptors in individual *Pieris brassicae* reared from the same egg batch, and on the same food. Each x represents the data for one individual. Responses to the deterrent strychnine are plotted against responses to the phagostimulant sinalbin. There is great variation in responsiveness of the deterrent cell, and independent variation in firing rate of the cell responding to sinalbin. The balance between input from positive (phagostimulatory) and negative (deterrent) inputs are different in different individuals, even though there is an overall correlation between the two across individuals. The broken line represents equal inputs, and the solid line is the fitted regression for the data (after Ma, 1972).

involving alterations in functioning of taste and odor receptors, and in properties of some relevant interneurons. In future studies, it may be possible to identify changes in receptors that are associated with different behavioral responses to food plants.

There is, as yet, no relevant work on the central nervous system, but there are some cases where intraspecific differences in food acceptance seem to have no correlates in terms of chemosensory inputs. This might provide at least negative evidence for a role of variation at higher neural levels. For example, in the silkworm, *Bombyx mori*, mutants have been found for broadened host range, yet the peripheral sensory system revealed no difference from that of the wild type. This supports the conclusion that changes in gustatory perception have arisen at the level of integration.

In conclusion, there is a shortage of information on exactly what changes physiologically when individuals differ in their behavior. Almost nothing is known of the genetics of neurophysiological factors relevant to the understanding of intraspecific, or even interspecific, genetic differences in host-choice behavior. A large field of endeavor awaits further research.

Further reading

Denno, R.F. and McClure, M.S. (eds.) 1983. *Variable Plants and Herbivores in Natural and Managed Systems*. Academic Press, New York.

Diehl, S.R. and Bush, G.L. 1984. An evolutionary and applied perspective of insect biotypes. A.Rev. Entomol. 29: 471–504.

Endler, J.A. 1986. *Natural Variation in the Wild*. Princeton University Press, New Jersey.

Futuyma, D.J. and Peterson, S.C. 1985. Genetic variation in the use of resources by insects. A.Rev.Entomol. 30: 217–238.

Hartl, D. 1981. *A Primer in Population Genetics*. Sinauer, Mass.

Huettel, M.D. (ed.) 1986. *Evolutionary Genetics of Invertebrate Behavior: Progress and Prospects*. Plenum, New York.

Papaj, D.R. and Rausher, M.D. 1983. Individual variation in host location by phytophagous insects. *In* Ahmad, S. (ed.) *Herbivorous Insects: Host Seeking Behavior and Mechanisms*. Academic Press, New York, pp. 77–124.

Roitberg, B.D. and Isman, M.B. (eds.) 1992. *Insect Chemical Ecology. An Evolutionary Approach*. Chapman & Hall, New York.

Singer, M.C. and Parmesan, C. 1993. Sources of variation in patterns of plant-insect association. Nature 361:251–253.

Strong, D.R., Lawton, J.H. and Southwood, R. 1984. *Insects on Plants*. Harvard University Press, Cambridge.

References (* indicates review)

Within population differences

Feder, J.L. and Bush, G.L. 1991. Genetic variation among apple and hawthorn host races of *Rhagoletis pomonella* across an ecological transition zone in the mid-western United States. Entomologia Exp.Appl. 59: 249–265.

Futuyma, D.J., Cort, R.P. and Noordwijk, I.van 1984. Adaptation to host plants in the fall cankerworm (*Alsophila pometaria*) and its bearing on the evolution of host affiliation in phytophagous insects. Amer.Nat. 123: 287–296.

Harrison, G.D. 1987. Host-plant discrimination and evolution of feeding preference in the Colorado potato beetle *Leptinotarsa decemlineata*. Physiol.Entomol. 12: 407–415.

Jaenike, J. 1985. Genetic and environmental determinants of food preference in *Drosophila tripunctata*. Evolution 39: 362–369.

Jaenike, J. 1986. Genetic complexity of host-selection behavior in *Drosophila*. Proc.Natl.Acad.Sci. 83: 2148–2151.

Jaenike, J.and Grimaldi, D. 1983. Genetic variation for host preference within and among populations of *Drosophila tripunctata*. Evolution 37: 1023–1033.

*Prokopy, R.J., Averill, A.L., Cooley, S.S., Roitberg, C.A. and Kallet, C. 1982. Variation in host acceptance pattern in apple maggot flies. *In* Visser, J.H. and Minks, A.K. (eds.) *Insect-Plant Relationships*. Pudoc, Wageningen, pp. 123–129.

Singer, M.C. 1983. Determinants of multiple host use by a phytophagous insect population. Evolution 37: 389–403.

Singer, M.C., Ng, D. and Thomas, C.D. 1988. Heritability of oviposition preference and its relationship to offspring performance within a single insect population. Evolution 42: 977–985.

Thompson, J.N. 1988a. Variation in preference and specificity in monophagous and oligophagous swallowtail butterflies. Evolution 42: 118–128.

Thompson, J.N. 1988b. Evolutionary genetics of oviposition preference in swallowtail butterflies. Evolution 42: 1223–1234.

Via, S. 1991. The genetic structure of host plant adaptation in a spatial patchwork: demographic variability among reciprocally transplanted pea aphid clones. Evolution 45: 827–852.

Waldvogel, M. and Gould, F. 1990. Variation in oviposition preference of *Heliothis virescens* in relation to macroevolutionary patterns of heliothine host range. Evolution 44: 1326–1337.

Wood, T.K. 1980. Divergence in the *Echinopa binotata* complex (Homoptera: Membracidae) effected by host plant adaptation. Evolution 36: 233–242

Biotypes

*Gould, F. 1983. Genetics of plant-herbivore systems: interactions between applied and basic study. *In* Denno, R.F. and McClure, M.S. (eds.) *Variable Plants and Animals in Natural and Managed Systems*. Academic Press, New York, pp. 599–653.

Gould, F. 1984. Role of behavior in the evolution of insect adaptation to insecticides and resistant host plants. Bull.Entomol.Soc.Am. 30: 33–41.

Gould, F. 1988. Evolutionary biology and genetically engineered crops. Bioscience 38: 26–33.

*Diehl, S.R. and Bush, G.L. 1984. An evolutionary and applied perspective of insect biotypes. A.Rev.Entomol. 29: 471–504.

*Via, S. 1990. Ecological genetics and host adaptation in herbivorous insects: the experimental study of evolution in natural and agricultural systems. A.Rev.Entomol. 35: 421–446.

*Wilhoit, L.R. 1992. Evolution of herbivore virulence to plant resistance: influence of variety mixtures. *In* Fritz, R.S. and Simms E.L. (eds.) *Plant Resistance to Herbivores and Pathogens*. Chicago University Press, Chicago, pp. 91–119.

Selection experiments

Claridge, M.F. and den Hollander, J. 1982. Virulence to rice cultivars and selection for virulence in populations of the brown planthopper *Nilaparvata lugens*. Entomologia Exp.Appl. 32: 213–221.

Fry, J.D. 1990. Trade-offs in fitness on different hosts: evidence from a selection experiment with a phytophagous mite. Am.Nat. 136: 569–580.

Gould, F. 1979. Rapid host range evolution in a population of the phytophagous mite *Tetranychus urticae*. Evolution 3: 791–802.

Wallin, A. 1988. The genetics of foraging behavior: artificial selection for food choice in larvae of the fruitfly, *Drosophila melanogaster*. Anim.Behav. 36: 106–114.

Wasserman, S.S. and Futuyma, D.J. 1981. Evolution of host plant utilization in laboratory populations of the southern cowpea weevil, *Callosobruchus maculatus* (Coleoptera: Bruchidae). Evolution 35: 605–617.

Population differences

*Alstad, D.N. and Edmunds, G.F. 1983. Adaptation, host specificity and gene flow in the black pineleaf scale. *In* Denno, R.F. and McClure, M.S. (eds.) *Variable Plants and Animals in Natural and Managed Systems*. Academic Press, New York, pp. 413–426.

*Fox, L.R. and Morrow, P.A. 1981. Specialization: species property or local phenomenon? Science 211: 889–893.

*Gould, F. 1988. Genetics of pairwise and multispecies plant-herbivore coevolution. *In* Spenser, K.C. (ed.) *Chemical Mediation of Coevolution*. Academic Press, San Diego, pp. 13–56.

Harrison, G.D. 1987. Host-plant discrimination and evolution of feeding preference in the Colorado potato beetle *Leptinotarsa decemlineata*. Physiol.Entomol. 12: 407–415.

Hsiao, T.H. 1978. Host plant adaptations among geographic populations of the Colorado potato beetle. Entomologia Exp.Appl. 24: 437–447.

Jones, R.E. 1987. Behavioural evolution in the cabbage butterfly (*Pieris rapae*). Oecologia 72: 69–76.

Knerer, G. and Atwood, C.E. 1973. Diprionid sawflies: polymorphism and speciation. Science 179: 1090–1099.

Leslie, J.F. and Dingle, H. 1983. A genetic basis of oviposition preference in the large milkweed bug, *Oncopeltus fasciatus*. Entomologia Exp.Appl. 34: 215–220.

Scriber, J.M. 1982. Food plants and speciation in the Papilio glaucus group. *In* Visser, J.H. and Minks, A.K. (eds.) *Insect-Plant Relationships*. Centre for Agricultural Publishing, Wageningen, pp. 307–314.

Scriber, J.M., Lederhouse, R.C. and Hagen, R.H. 1991. Foodplants and evolution with *Papilio glaucus* and *Papilio troilus* species groups. *In* Price, P.W., Lewinsohn, T.M., Fernandez, G.W. and Benson, W.W. (eds.) *Plant-Animal Interactions*. Wiley Interscience, New York, pp. 341–374.

Singer, M.C. 1971. Evolution of food plant preference in the butterfly *Euphydryas editha*. Evolution 25: 383–389.

Singer, M.C. 1993. Behavioral constraints on the evolutionary expansion of insect diet breadth. In Real, L. (ed.) *Behavioral Mechanisms in Evolutionary Ecology*. Univ. Chicago Press.

Singer, M.C., Ng, D. and Moore, R.A. 1991. Genetic variation in oviposition preference between butterfly populations. J.Insect Behav. 4: 531–536.

Tabashnik, B.E. 1983. Host range evolution: the shift from native legume hosts to alfalfa by the butterfly *Colias philodice*. Evolution 37: 150–162.

*Tabashnik, B.E. 1986. Evolution of host plant utilization in *Colias* butterflies. *In* Huettel, M.D. (ed.) *Evolutionary Genetics of Invertebrate Behavior: Progress and Prospects*. Plenum Press, New York, pp. 173–184.

Tavormina, S.J. 1982. Sympatric genetic divergence in the leaf-mining insect *Liriomyza brassicae* (Diptera: Agromyzidae). Evolution 36: 523–534.

Thompson, J.N. 1988. Evolutionary genetics of oviposition preference in swallowtail butterflies. Evolution 42: 1223–1234.

Mechanisms

Ayyub, C. Paranjape, J., Rodrigues, V. and Siddiqi, O. 1990. Genetics of olfactory behavior in *Drosophila melanogaster*. J.Neurogenet. 6: 243–262.

Ishikawa, S., Tazima, Y. and Hirao, T. 1963. Responses of the chemoreceptors of maxillary sensory hairs in a "non-preference" mutant of the silkworm. J.Sericult.Sci.Jpn. 32: 125–129.

Harrison, G.D. and Mitchell, B.K. 1988. Host-plant acceptance by geographic populations of the Colorado potato beetle, *Leptinotarsa decemlineata*. Role of solanaceous alkaloids as sensory deterrents. J.Chem.Ecol. 14: 777–788.

Ma, W.-C. 1972. Dynamics of feeding responses in *Pieris brassicae* Linn. as a function of chemosensory input: a behavioural, ultrastructural and electrophysiological study. Meded.Landbouwhogeschool Wageningen, 72, 11.

Venard, R. and Stocker R.F. 1991. Behavioral and electroantennogram stimulation in *lozenge*: a *Drosophila* mutant lacking antennal basiconic sensilla. J.Insect Behav. 4: 683–705.

*Visser, J.H. 1983. Differential sensory perceptions of plant compounds by insects. *In* Hedin, P.A. (ed.) *Plant Resistance to Insects*. ACS Symposium Series 208, American Chemical Society, Washington D.C., pp. 215–230.

Wieczorek, H. 1976. The glycoside receptor of the larvae of *Mamestra brassicae* L. (Lepidoptera, Noctuidae). J.Comp.Physiol. 106: 153–176.

8

Evolution of Host Range

In previous chapters, we addressed the mechanisms of host choice. This chapter deals with questions of the ultimate or functional bases of different host-choice patterns. What determines diet breadth and why are so many phytophagous insects relative specialists? The patterns of feeding illustrated in Chapter 1 indicate the degree to which narrow diets are common in different phytophagous insect groups. Yet, there are clearly advantages in being able to feed on many items; insect species with generalized feeding abilities can have a broad geographical range, many generations each year, and an almost certain availability of food at any time. Individuals may also take advantage of being able to select among foods to balance the nutrient intake. At face value, fitness should be greater for polyphagous species, but despite this, most species are specialists. Therefore, interest has focused on the reasons for specialization. It is generally thought that there are many factors involved in the evolution of, and maintenance of specialization, and that no one of them can be singled out as predominant. In addition, it seems likely that the underlying reasons may differ among insect groups.

In some cases, the factors that seem to be related to the evolution of specialized diets may be such that they increase the likelihood of specialization rather than providing direct selective pressures for it. An example is availability of the food resource. Specialization on a particular plant species is more likely to evolve if the plant is abundant and predictably available, because individual insects are more likely to encounter it. Availability is not a selective pressure itself but provides the opportunity for selection for narrow host range to occur.

Sometimes it is difficult to say whether a factor caused specialization, or evolved later as a result of specialization. For example, if specialists show a low level of tolerance to various plant secondary metabolites, it is usually unclear whether their inabilities are specifically associated with evolution of specialized feeding, or whether, after evolving a narrow diet, the ability to deal with chemicals that are not encountered was subsequently lost.

Finally, the origin of specialization may have had causes that are different from selective factors or advantages that now maintain it. If an insect species becomes a specialist on a plant for some ecological reason, and subsequently becomes highly cryptic on that plant, it is likely to have an advantage over a more generalized species with respect to predation. In this scenario, the specialized morphology would be important in maintenance of specialization, since host-specific crypsis is less likely to provide protection on alternative plant types. But it is much less likely that a well developed crypsis would have predated the specialized behavior.

In addition to specific factors that relate to the evolution of specialized diets, it is clear that phylogenetic history of insect groups is important. Different lineages appear to show very different tendencies to be either broad or narrow in their diets. However, the bases of these different tendencies inevitably lie with one or more of the factors that will be discussed. The consideration of phylogeny will therefore be addressed in the context of the different correlates associated with narrow diet breadth.

Table 8.1 presents the variety of factors that have been associated one way or another with specialized feeding habits, and provides a framework for this chapter. We will address these different factors, and present relevant published

Table 8.1. Correlates of restricted diet breadth

Enabling factors for specialization
1. Resource availability—availability of a suitable plant species in a predictable manner
   a. availability in space, meaning large numbers, large areas, and/or large species
   b. availability over time, both ecological and evolutionary
   c. availability in relation to insect size
2. Competition between species may lead to division of resources among them
3. Behavioral patterns unassociated with feeding may play a role as when thermal contraints prevent the use of most plants in a forest habitat
4. Accumulation of secondary compounds in plants may limit acceptability of many plant species
5. Limitations in sensory processing. This may mean that efficiency of foraging is greatest when choices are few

Post-specialization factors maintaining limited diets
1. Morphology—adaptations to specific host
2. Postingestive physiology—adaptations to specific plants, with subsequent loss of generalized abilities
3. Sensory neural changes and associated behavioral adaptation, such as rejection responses to increasing numbers of plant secondary metabolites.
4. Reproductive behavior. The host may become the point where the sexes meet and can provide a benefit in terms of a mechanism for mate finding.
5. Phenological matching between insect and host may limit diet breadth
6. Specific behaviors associated with a particular host that can enhance avoidance of mortality factors

data to illustrate whether, and to what extent, they seem to be important. We conclude with some attempt at bringing the different factors together, and provide some overall theme to the question of why so many phytophagous insect species are relative specialists in their feeding habits.

8.1 Factors relating to scale

The average insect body length is approximately 4mm. While vertebrate herbivores rarely would have adequate quantities of food if they were restricted to one or two plant species, insects, being so small, at least have the capability of being specialists. Furthermore, total overall resource needs obviously vary with the size of the herbivore. Among vertebrates, the few specialists are those for which enormous regions are dominated by single plant genera. For example, eucalypts in Australia have allowed the evolution of the specialist koalas, and bamboo in China has allowed the evolution of the specialized feeding habits of the pandas. The question of body size within the insects, and resource levels available, are discussed in this section.

8.1.1 Insect size

Is there a pattern within the phytophagous insects of larger insects being less specific than small ones? Some studies have been undertaken within the Lepidoptera. British species were divided into micro- and macroforms and the host ranges examined. The data are shown in Fig. 8.1 and clearly indicate a pattern of

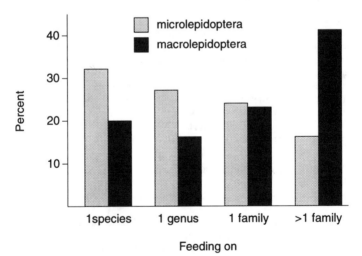

Figure 8.1. Relative diet breadths of macro- and microlepidoptera in Britain. Bigger percentages of microlepidoptera feed within one plant species or genus, while larger percentages of macrolepidoptera feed on plants in more than one family (after Gaston and Reavey, 1989).

broader diets in the macrolepidoptera compared with the microlepidoptera. In a study with the larger species of Lepidoptera of New York state, it was found that overall, diet breadth was positively and significantly correlated with size. In further studies, microlepidoptera were divided into internal and external feeders. The internal feeders had an average wing span of 8 mm as adults, while the external feeders had an average wing span of 16 mm. Internal feeders were notably more specific with 72% feeding within a genus of plants, while the external feeders had less than 50% of species feeding within a genus.

In considering the Lepidoptera, there is still the question of whether being small led to specialization or whether, being a specialist provides a basis for becoming smaller. Furthermore, the micro- and macrolepidoptera have little overlap taxonomically so that there may be phylogenetic differences that predispose different groups to be narrow or broad in their diets. Only a detailed study of phylogenetic history will help to resolve the question of which comes first, narrow host range or small size.

8.1.2 Resource availability

It has often been suggested that when a plant species is extremely abundant and totally reliable, encounter rates of insects will be relatively high and, eventually, individuals of apparently unadapted species will probably accept the plant, survive and reproduce. In situations where the abundant plant becomes dominant in a region, fitness of the herbivores may be greatest if they restrict themselves to that plant. Hence there is a logical argument that great abundance of a particular plant species could, in itself, lead to specialization by some herbivore species. What is the evidence?

The number of different phytophagous insect species is strongly correlated with relative abundance of plants on which they feed. Fig. 8.2 demonstrates the pattern in two systems: insect herbivores on trees in Britain, and lepidopterous leaf miners on oaks in North America. Although this says nothing about diet breadth, it does indicate the importance of resource availability in the accumulation of species, and highlights the fact that encounter rate probably will, in species having a degree of appropriate variability, eventually lead to acceptance. This could then be the first step in a series of changes leading to narrow diets. That specialization may subsequently occur is suggested by the fact that trees which have been present in Britain for longer periods of time have more species of insects feeding upon them, and a greater proportion of specialists, than tree species more recently introduced.

Colonization of soybean in North America provides an example of what happens in a shorter time scale. Soybean was introduced to North America from China in about 1910 and has become extremely widespread as a cultivated crop (40 million hectares in 1980). It became colonized, first by polyphagous species, then by species adapted to other legumes, and finally by specialist species from

experience could accumulate faster?

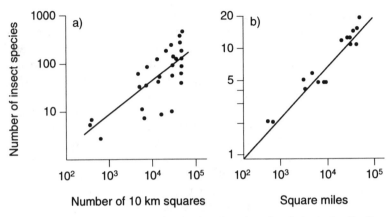

Figure 8.2. Numbers of insect species found on plants in relation to the distribution area of plants. **a)** Numbers of species of all phytophagous insects on British trees in relation to the abundance of the tree species. Abundance is based on the number of 10km squares in which each genus of tree occurs (after Strong, Lawton and Southwood, 1984). **b)** Numbers of leaf-mining lepidopteran species in relation to area covered by species of oaks in North America (after Opler, 1974).

other plant families. Some of the specialists now prefer soybean to their original hosts. There is also a suggestion that some of the species in the broader diet categories are becoming more restricted. In its native habitat in China, a majority of insect herbivores on soybean are specialists. The evolution of soybean-feeding herbivores in North America will probably continue, but it remains to be seen whether the degree of specialization already observed will continue to evolve.

Other evidence that changes in availability of particular plant taxa may lead to increased specialization in host choice comes from the study of present-day host associations of grasshoppers. The acridids are known to have an origin and history of polyphagy, and the majority of species today maintain a broad diet. However, where grasslands evolved and covered large areas, considerable numbers of grasshopper species became specialists on grasses. Thus in Africa nearly 50% of the 1400 or so grasshopper species are graminivores. In North America, where grasslands have covered less of the continent and are more restricted to temperate zones, there is a smaller proportion of grass specialists. Overall, becoming specific on grasses appears to have occurred independently eight different times within the grasshoppers.

In the deserts of southwest North America, where grasses have become ephemeral and unpredictable, two species from the grasshopper subfamily Gomphocerinae, *Bootettix argentatus* and *Ligurotettix coquilletti*, have become specialists on the widespread, dominant and totally reliable creosote bush (family Zygophyllaceae). Since the Gomphocerinae are otherwise grass specialists, the indication is that these species have changed their diets and become specialists on a new plant in response to resource availability.

Although resource abundance alone is thought to give rise to specialized feeding habits, this tendency may be enhanced by the limited availability or reliability of the previous hosts. In the case of grasshoppers specializing on creosote bush, this may have been a significant factor in the shift to the new host. The grasses which were the original hosts, became more and more unreliable as a resource as desertification increased, while the availability of creosote bush increased.

Unreliability of plants has been shown to matter in other more subtle ways. In the Mediterranean region, which is the center of species diversity for cruciferous plants, some crucifer species are abundant every year, while others are abundant in some years and not others. These crucifers are used by a variety of species of pierid butterflies. Diet breadth of butterflies was found to be greatest with species that utilized plants having variable abundance (Fig. 8.3). On the other hand, butterflies that utilized plant species which were uniformly abundant in different years tended to have a narrower host range. In other words, where host plants were shown to be more variable in availability from year to year,

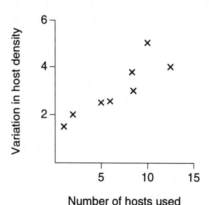

Figure 8.3. Diet breadth of pierid butterflies in relation to host reliability in North Africa. Each point shows the number of host plants used by one butterfly species for oviposition, in relation to variation in host density. The variation in host density was measured as density in year 1 divided by density in year 2 + 1. Based on a three year study at 49 sites (after Courtney and Chew, 1987).

butterflies had broader diets; where hosts were consistently available, butterfly species had narrower diets.

The fact that monophagous insect species occur on most plant species, independent of their abundance, makes the resource availability hypothesis seem rather limited in its applicability, but without knowing the history of abundance of host plants, it is difficult to put weight on this. In Hawaii, with a relatively short geological history, it is notable that where insect groups have been studied, the monophagous ones utilize common plants. For example, it has been clearly shown among the numerous monophagous species of *Drosophila*, that, almost without exception, they are restricted to the abundant species of plants.

The data on the importance of resource availability are not conclusive, but do suggest quite strongly, that if plant species are extremely abundant and reliable, insects are able to specialize on them, and often do so. If plant species are less reliable, specialization may be less likely to evolve. In any case, resource availability would be expected to influence diet breadth, and perhaps could be thought of as a possible prerequisite for the evolution of specialization.

8.1.3 *Plant phenological factors*

Plants vary in their suitability for insect herbivores during development (see Chapter 2). Not only does foliage quality change over time, but the availability and quality of flowers, fruit and seeds is usually temporally very restricted. Furthermore, many herbaceous plants are available for a limited period of the year. The timing of this period varies with habitat and with plant taxon. In deserts, it may be associated with rain; at high altitudes, with summer temperatures. Different plant families tend to have different phenologies, flowering at different times. For example, a high proportion of species of Asteraceae flower late in the summer or fall. For whatever reason, the availability of host-plant material is often temporally very restricted. If an insect species becomes adapted to the use of one such plant there will be temporal constraints on the availability of others. This will automatically have the effect of reducing host range.

A comparative study of two lycaenid butterflies in the Rocky Mountains provides an example of how such constraints might operate. One species, *Glaucopsyche lygdamus*, uses a wide variety of leguminous plant species as hosts, while the other, *Plebejus icarioides*, is specific to lupines. *G. lygdamus* oviposits on flowers and the larvae feed on pollen. The species of host plants used had available inflorescences at very different times of the season, and the species composition of suitable hosts also varied from year to year. This meant that patterns of host species used varied, depending on their availability at the appropriate phenological stage. If the butterfly were restricted to one host plant, it would face the risk of not having its host in some years. Thus, phenological variation among hosts from year to year, probably prevents this species from specializing on any one. On the other hand, *P. icarioides*, utilizes old leaves when it first hatches. Because these are continuously available, this species is able to be a relative specialist. In this comparative study, it appears that the plant part utilized influenced the range of hosts used, because the more ephemeral tissue was relatively less predictable.

Temporal limitations in specific food availability have been studied in a few other cases. For example, the synchrony of insect hatching with bud burst in trees, and the subsequent availability of young leaves for newly hatched insect larvae, has been cited as a major problem for tree-feeding caterpillars. Commonly, eggs are laid on twigs of trees after leaf fall, and, in the spring, insects that hatch too soon have no food, while those that hatch too late have leaves that are older, tougher and often nutritionally inadequate for development. In the fall cankerworm, *Alsophila pometaria*, studied in New York State, different genotypes predominate on different tree species. Furthermore, there are consistent differences among the genotypes in hatching times in spring. These differences correlate with the approximate times of bud burst in the hosts, and it seems likely that the phenological differences contribute to the specific genotype-host associations. It may be that this differentiation in genotype frequency on different hosts represents an early stage in evolution of several species with different specialized host associations from a single polyphagous species.

Plant phenology can thus be significant in relation to host-plant range. It is, like resource availability generally, a factor that may allow or constrain plant use. It is not clear, however, to what extent insect herbivores have adapted to host phenology following host restriction, or to what extent plant phenological characteristics have determined diet breadth. Further, the differential use of plant species according to phenological differences among them may have its origin in competitive interactions (see below).

8.2 Interactions among insects

Interactions within and among herbivorous insect species have been considered important in relation to host range, while it has also been suggested that generalist predators may provide selection pressure for specialized diets. The case for interactions influencing host range is briefly described in this section.

8.2.1 Competition

One of the theories associated with diet breadth is that narrow host ranges result from interspecific competition. The basis for this concept is that species may maximize fitness by developing means of monopolizing a particular host plant and excluding other species. In this way, the plant resources (usually species) would become split up among the different herbivores which are then, by necessity, relative specialists. Thus the relationship of competition to diet breadth is that species take up specific niches in which they become particularly competent, and thus outcompete others.

The first requirement in determining that interspecific competition is relevant to the development of host specificity, is to demonstrate that such competition occurs. Certain authors have argued that competition is rare among insects, and there are several studies with phytophagous insect guilds that provide no evidence at all of competition. It is not surprising therefore, that the importance of competition is debated. A recent review by Damman indicates that, of 99 studies, 58 gave evidence of competitive interactions, meaning that there was a negative effect on one or more of the interacting species.

While competition may be more common or conspicuous in vertebrates or in arthropod predators, there is clear evidence that competition can occur among phytophagous insects. Disagreements concerning the relative importance of competition in insect herbivores, partly result from the fact that, in ecology, some look for processes that currently structure populations, while, in evolutionary biology, competition is seen more as having significance over evolutionary time. However, in an equilibrium situation, one might expect to see relatively little competition, even if it has been a major force in the past. The significant periods of competitive interactions that tend to minimize niche overlap may be quite short in evolutionary terms. Therefore, they are most likely to be found in situations where there has been biological disturbance, and many of the examples of competition among herbivores come from situations where plants or insects have been introduced into new regions.

One example of a recent biological disturbance concerns scale insects on eastern hemlock, *Tsuga canadensis*. Two species of scale insects, *Fiorinia externa* and *Tsugaspidiotus tsugae*, were apparently introduced into Canada from Japan in about 1910. Both feed on mesophyll cells of the youngest needles in the lower part of the crown of *T.canadensis* trees. Thus, they occupy the same feeding niche. Experimental studies from 1978 to 1981 demonstrated that they do compete, while field surveys similarly indicated competition (Fig. 8.4). When single trees were occupied by only one of the species, population densities of the two species were similar, but when both were present on one tree, one was always at higher density than the other. In some trees, one species outcompeted the other completely, so that only one remained.

An example of competition on a recently introduced plant involves three different species of *Rhagoletis* flies in Wisconsin. These have shifted from their native hosts to a single introduced host, sour cherry. All three simultaneously infest the fruit and none recognizes the others' oviposition marking pheromone, resulting in a very high degree of interspecific competition. It is common for all individuals to die in a fruit due to such competition. It remains to be seen, but the situation appears to be one in which, sooner or later, selection will favor flies restricting themselves to single hosts, either their original native plants or the introduced sour cherry. In such a manner, competition would be reduced.

There are also examples of intraspecific competition. Oviposition marking pheromones are common among herbivores and very often these deter conspecifics.

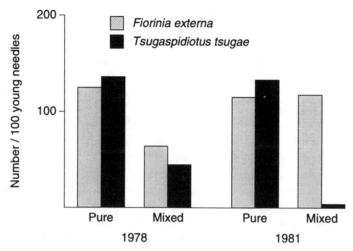

Figure 8.4. An example of competitive interactions among two species of scale insect, *Fiorinia externa* and *Tsugaspidiotus tsugae*, that inhabit hemlock trees, and utilize the same parts of the tree and the same types of needles. Average density of each species is shown for sampling periods in 1978 and 1981. In mixed stands, *F. externa* gives evidence of outcompeting *T. tsugae* (after McClure, 1980).

This may be evidence of intraspecific competition. It is conceivable that, under some circumstances, this could lead to reduction in host range. Such a situation may occur when a species using two or more hosts splits up into two separate populations using separate hosts during the process of sympatric speciation (see Chapter 7). More commonly, however, intraspecific competition is seen as a pressure for alternative host use such that increasing diet breadth is favored.

In conclusion, competitive interactions do occur, and it seems reasonable to believe that they bring about situations in which specialization might evolve. The limited number of current examples of competition is inevitable, if what we see is the result of competition in the past.

8.2.2 Sexual interactions

Sexual encounters often occur on host plants. This has led to the theory that narrow host range is important in sexual selection and mate finding. That is, narrow host range has specific fitness benefits in that mate finding on specific hosts will be easier than mate finding on diverse hosts. The importance of mate finding, as a reason for narrow host range has been termed the "sexual rendezvous hypothesis." It is most reasonable as a benefit of narrow host range when the host plants are relatively scattered. An example comes from work on hummingbird flower mites. These very small herbivores are mostly monophagous and live in corollas of flowers in the tropics; they have abundant resources, and there is no evidence of competition. They are carried from flower to flower by hummingbirds and disembark only at their own host in response to the odor of the nectar. Because flower mites mate only in their hosts, and they mate many times, straying from the normal host will lead to fewer descendants. Therefore, selection arising from differential mating success may narrow or focus host range and intensify host fidelity.

Among aphids, evidence suggests that mate finding may have been a factor determining host range. In those species that alternate between the primary host in winter and the secondary host in summer (heteroecious), mating takes place on the winter host and the range of hosts for this aspect of the life history is narrow and very conservative. From the few available studies, it appears that over 99% of individuals returning to the winter host never get there, and the problem of finding a mate would surely be compounded if there were not strict limits on host use. Heteroecious species generally have very few primary hosts, but often (about 50%) have polyphagous summer parthenogenetic forms. In other words, the sexual generation shows a host restriction, while the asexual generations seem to be polyphagous. A simple interpretation would suggest that there is some particular value in conserving the narrow host range in the generation where males and females interact, but that in the parthenogenic generations sexual constraints are absent and broader diets predominate.

If we remain with the complexities of aphids, some species remain throughout the year on the herbaceous host plants that are used in summer (autoecious), and go through the sexual generation on them; they do not migrate to a winter host. Among such species only 11% feed on more than one genus. That is, there is considerable specialization on these hosts (Fig.8.5). This contrasts with the species that use similar hosts solely for their parthenogenetic generations, and, once again, suggests that the presence of a sexual generation on these plants might be related to a restriction of host range.

In many species of herbivorous insects, males as well as females are attracted to the odor of the host plant. In others, males are most attracted to females in the presence of the host odor. For example, the males of the leek moth, *Acrolepiopsis assectella*, are attracted to females in the presence of the odor of leeks. In such cases, it may be suspected that the mating rendezvous hypothesis is relevant, and that an advantage of the narrow host range is the relative ease with which males can find females.

In conclusion, the few known cases of host-plant influences in mate finding indicate that specialization could be a sexual benefit. The benefits have probably to be seen as secondary to the origin of specialization, but perhaps very important in maintenance of specialization, at least in those groups where it clearly occurs.

8.2.3 Predators and parasites

L. Brower first suggested in 1958 that predation could be an important factor in evolution of narrow diets of phytophagous insects. His scenario was that two

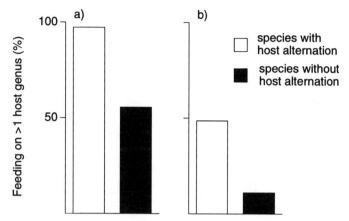

Figure 8.5. Diet breadth in aphids. Aphid species which alternate between a winter host and a summer host, have many host species in the summer. On the summer hosts they are parthenogenetic. At the end of summer winged forms migrate to the winter host where mating occurs. Aphid species that go through both the parthenogenetic and sexual generations on the same host plants have host ranges that are narrower than those of aphids spending only the parthenogenetic generations on similar hosts. **a)** grass-feeding species (after Ward, 1992). **b)** European Aphidinae (after Eastop, 1973).

plant species, A and B, are occupied by each of two lepidopteran phenotypes, 1 and 2. Caterpillar 1 is better protected (i.e., more cryptic) on A and caterpillar 2 is better protected on B. Persistent predation by birds would then lead to the restriction of 1 to plant A, and 2 to plant B. In other words, predation would cause the evolution of more specific host choice.

Thirty years later, the theme has surfaced again, and recent experiments demonstrate clearly that in several systems generalist natural enemies are important in the maintenance of narrow host range. Experiments were first carried out with vespid wasps on 28 species pairs of caterpillars in a seminatural situation. In these experiments, the wasps foraged among the plants which were naturally fed upon by the caterpillars. Each species pair consisted of a specialist and a generalist. Generalists were significantly more vulnerable (Fig. 8.6). The specialists appeared to benefit from a wide variety of behavioral, chemical and morphological characters. Using techniques appropriate to the different systems, a similar pattern was demonstrated among ants and small caterpillar prey, and among coccinellids and aphids. In summary, these experiments showed unequivocally that being a specialist is an advantage. In other words, generalist predators have a greater impact on insect herbivores with greater diet breadths, so that, other things being equal, one can say that generalist predators should be important in at least maintaining narrow host ranges of insect herbivores.

A variety of studies on individual plants or insect pairs tend to support the

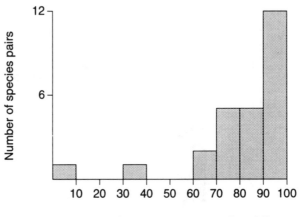

Figure 8.6. Predation by generalist predators on caterpillars with different diet breadths. The wasps, *Mischocyttarus flavitarsus*, were foraging naturally in a large greenhouse where they were regularly fed with dead potato tuber moth caterpillars. In experiments, they were allowed to forage on caterpillars of other species which were feeding on their natural hosts. Two species of caterpillars, contrasted in their diet breadth, were available in any one experiment. Wasps killed caterpillars and took them away to provision their nests, and these events were continuously observed until approximately half of either caterpillar species had been taken. Data are presented as the percent of all insects taken in an experiment that were the generalists in the pair. If the figure is 50% both were taken equally, if 100% then only the generalist was taken. These data show a much greater vulnerability of caterpillars with a wide host range compared with the specialists (after Bernays, 1989).

idea that predation is a more important source of mortality among generalists than specialists. *Phratora vitellinae* beetles utilize willows. It was found that they strongly prefer *Salix fragilis* over other species in Scandinavia, even though this is not the best plant for development. All larvae raised on this plant survived when exposed to coccinellid beetle predators, but not when reared on other willow species. The insects sequester salicylates from willow and *S.fragilis* has the highest levels. Beetles reared on this willow have high levels of defensive secretions containing salicyl aldehyde. Similar studies have been made on other species that sequester defensive chemicals from their hosts. Once again, it seems that the benefit of being specialized is that mortality from predators is reduced. Perhaps, therefore, predators are important in maintenance of specialization.

There are isolated examples throughout the literature supporting the notion that parasites and predators may be relevant in relation to host-plant range of herbivores. For instance, several generalist aphidiid parasites were found to locate, or feed more readily on, generalist aphids compared with specialist aphids. In studies with bracken fern in Britain, specialist herbivores were found to be unaffected by generalist predatory ants that visit the nectaries on these plants, while generalist herbivores were fiercely attacked. In studies of two species of the aphid genus *Uroleucon*, on *Solidago* in North America, there were found to be significant differences in survival from attack by natural enemies, the specialist being at an advantage. In a final example, *Drosophila* species have apparently been influenced in the hosts they use by certain nematode parasites. Feeding on amanitin-containing mushrooms reduces the survivorship of parasitic nematodes in flies, and increases survivorship of the flies in the presence of nematodes. Other mushrooms give better survival of the flies in the absence of nematodes, but flies prefer those containing amanitin, and consequently obtain protection from the nematode parasites.

In summary, there is reason to believe that generalist natural enemies have an important role in maintenance of narrow host range. In terms of the evolution of narrow host range, predation probably acts in concert with other factors, especially resource availability. A theoretical example of such a situation is as follows:

(a) A generalist herbivore finds itself in a new environment with plants available in different proportions. Of four acceptable plant species, A,B,C,D, availability is in the proportions 10:5:3:1.

(b) The herbivore can feed on all four species equally, so that the predicted feeding on the four species will be in proportion to their availability; namely 10:5:3:1.

(c) Since A is most abundant, most time spent feeding will be on A. In this case, insects will benefit if they are relatively more cryptic on A than on B,C or D. There will be selection for individuals less conspicuous on A. If they suffer less predation on A, then there is likely to be selection towards a preference for A. In other words, the role of predators will be to enhance the preference of insects for the most abundant plant.

The conclusion is that generalist predators and parasitoids will be important in maintenance of narrow host ranges, and in some cases are likely to have been involved in the evolution of narrow host range.

8.3 Plant chemistry

It has been known for a long time that, in many groups of insects, there is a tendency for related species to feed on related plants with similar chemistries.

For example, among pierid butterflies, the Pierini feed on crucifers, while the Coliadinae feed on legumes. Many similar examples could be given. Even among higher taxa, there is evidence that the more ancient insect families are associated most with phylogenetically ancient vascular plant groups. Thus the family Coccidae (scale insects), which has a very early origin, is associated mainly with the early plant groups such as conifers. On the other hand, the Tephritidae (fruit flies), a relatively very young family, has a very high proportion of species associated with more recently evolved families of plants. Thus on the average, modern phytophagous groups have adopted modern plant groups as hosts while older groups maintained, in part, an ancient association.

A recent approach to understanding these relationships has been to examine phylogenies of both the plant and the insect clades. An example is the phylogenetic analysis of *Phyllobrotica* leaf beetles and their host plants in the mint family. There is a striking similarity in the two phylogenetic trees (Fig. 8.7) indicating a long history of close association, and the maintenance of specialized behavioral patterns. Since chemistry is of critical importance in the host-selection behavior we see today, and there are strong chemical similarities among the mint species, it seems probable that the beetles coevolved with, or tracked the evolution of, the mints as the plant chemistry diversified. From such trees it is clear that specialists have given rise to more specialists, and the apparent preponderance of this pattern has given rise to the idea that the large number of relative specialists

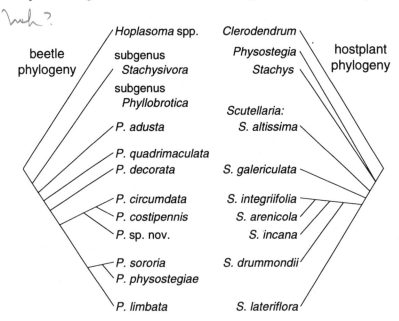

Figure 8.7. A comparison of phylogeny estimates for *Phyllobrotica* leaf beetles and their host plants (after Mitter and Farrell, 1991).

amongst the insects may have arisen because of a greater number of speciation events among specialists. In other words, it is possible that the relative abundance of specialists today is a result of more rapid speciation among specialists than among generalists.

In this section we examine the variety of influences that plant chemistry may have had in the evolution or maintenance of narrow host range.

8.3.1 Behavior

We have pointed to the important role of deterrent chemicals as determinants of host range in Chapter 4, but the role of deterrence in the evolution of specificity may have arisen in different ways. A chemical that is deterrent may have become so for several different reasons. It may be detrimental if it is ingested and this makes it unacceptable. In this case, one could expect the chemistry to be a direct selective factor for reduced host range. For example, among several species of host plants some may evolve new compounds which are deleterious to individuals of an insect species when ingested. Selection will favor insects that reject these plants and accept only the alternative species. Alternatively, the chemical may signify, in an evolutionary sense, a plant on which fitness is reduced due to various ecological factors. In this case, a plant species may be rejected for reasons unrelated to the postingestive effects of the chemical. For example, if a species of insect feeds on plant species A and B, but greater predation occurs on B, then female insects may be selected which are deterred by chemicals in plant B. In this way, chemicals may become deterrents and effectively reduce the host range. Thus, chemistry combined with other fitness parameters leads to narrower host range. A third possibility is that chemicals are inherently deterrent because of the properties of the sensory system alone. In this case, the effect is to prevent extension of the host range even though there may be no fitness cost.

Without information on the details of behavior and physiology it is impossible to even suggest how common these different patterns may have been. We do know that there are many harmless deterrents. We also know that the chemistry of plants provides constraints on specialists, in that they are strongly deterred by nonhost compounds and often specifically stimulated by host compounds. These constraints probably determine the tracking of plant evolution by insect groups, and explain the parallel nature of plant and insect clades. The commitment of an insect taxon to a plant lineage may thus reduce the likelihood of reversal to more polyphagous habits. In other words, plant chemistry is probably important in the maintenance of narrow host ranges. It is more difficult to assess its importance in the origin of host specificity.

8.3.2 Digestive and toxicological factors

One might predict that specialists would be able to derive nutritional benefits from adapting to specific diets. One of the hypotheses to explain narrow diets in phytophagous insects is that adaptation to a specific food is associated with a reduced ability to deal with many different foods. Since 1970, workers have been investigating this in a variety of different ways. For example, studies have sought to identify whether specialists are in some way better at digestion and detoxification of their specific hosts than are generalists. Ten separate studies in the last two decades have given the following results: two demonstrated advantages in the more specific habits, six found no differences, and two found that generalist patterns of food intake led to fitness advantages. Clearly, at this level, the evidence does not enable us to generalize about postingestive benefits of being a specialist.

At the intraspecific level, the approach is similar; genotypes that have a very narrow host use should show greater fitness on the appropriate hosts than genotypes with broader host-use patterns if plant chemistry is of signal importance. Further, if a genotype becomes particularly competent at the use of one resource, and at the same time loses an ability to be as efficient on alternative resources, there would be strong evidence for the idea that limited host range should be selected for if there are no other important costs.

In one study with the caterpillars of *Alsophila pometaria*, ten clones were tested on four different hosts. It was found that the clone from one host plant grew better on its own host than on other hosts, but this pattern was not found with the other clones. The occurrence of even one such example indicates the possibility that if populations become especially competent at using one host, with the concomitant loss of competence on other hosts, then selection would favor narrow host ranges.

In studies with the phytophagous mite, *Tetranychus urticae*, there was some evidence of such changes. Lines were selected on different host plants in the laboratory. Some lines were kept on lima beans, a suitable host, and other lines were selected on tomato, on which there is normally high mortality and low fecundity. After ten generations the tomato-selected lines were better at using tomato than were the bean lines. Some tomato lines were then reselected on bean (reversion lines) and retested at intervals on tomato. Fig. 8.8 shows that as they became adapted once again to bean, they lost some of the ability to deal with tomato. This is evidence that, by becoming a specialist on one host, the ability to use an alternative host may be reduced. These negative genetic correlations in performance across host species, or tradeoffs, as they are termed, may therefore indicate that it is advantageous to become a specialist, rather than maintaining an ability to use many hosts.

While such tradeoffs have been hard to find in relation to digestive/toxicological ability, it may require a longer time frame to see losses of competence on previous hosts. At present the conclusion must be that there is some limited evidence in favor of physiological tradeoffs as a basis for the evolution of specialization. The limitations in host-use capabilities of many current specialists, however, are most likely to have been a result of specialization; following the evolution of

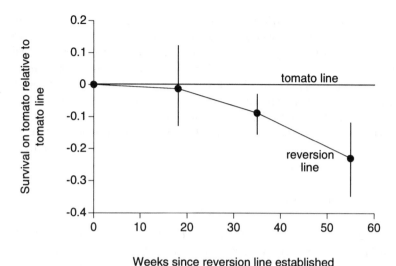

Figure 8.8. An experiment to determine if selection for survival on one host plant reduces survival on other plant species. The mite, *Tetranychus urticae*, was selected on tomato in the laboratory. From this tomato line, mites were taken and allowed to adapt to lima bean. These so-called reversion lines were tested at intervals for their ability to survive on tomato, and this was compared with survival on the original tomato line. The figure shows that performance on tomato became poorer with time since the reversion line was established. This suggests that adaptation to lima bean made individuals less able to deal with tomato. Vertical bars are 95% confidence limits (after Fry, 1990).

specialized host use, certain capabilities to use alternative foods are subsequently lost. This would mean that the limitations of specialist herbivores would be a result rather than a cause of being specialized. However, the adaptations that do develop might, at least, be important in the maintenance of a narrow host range.

A long history of association between an insect taxon and a particular group of plants seems generally likely to cause a progressive accumulation of adaptations that would tend to commit an insect lineage to the plant group. This has been suggested above at the behavioral level, and it is likely also at the post-ingestive level. Among grasshoppers that have become restricted to feeding on grasses, toxic plant secondary compounds are apparently more potent than in polyphagous species. The important point in this context is that grasses are noted for being impoverished in terms of such secondary metabolites. Thus the inability of such grass-specialists to tolerate a variety of noxious secondary chemicals found in broad-leaved plants is most likely to be derived, following the evolution of specialized feeding behavior.

To what extent post-ingestive physiology is important in host-plant specificity, we are not usually able to say. Certainly, however, evolutionary history has a major impact on what a lineage of insects can or will do. Extreme antiquity of certain insect-host associations has been documented, and among specialists extraordinary abilities to deal with noxious chemicals in the host plant have been recorded. Such ability may involve close evolutionary interactions with the plant, increasing commitment to that plant, and decreasing competence on other plants. There is some limited evidence that competence on one plant is sometimes linked to reduced competence on others. On balance, one could expect that plant chemistry has had impact on the evolution of specialization, but more particularly is important in the maintenance of specialization.

8.4 Insect characteristics

Some features of insects may be different from those of other animals from which theories of food selection have been derived. For example, optimal foraging theory as applied to hummingbirds can have little relevance in a rather immobile caterpillar. We here consider some of the possible insect features that may be of specific relevance to host-plant range.

8.4.1 Life histories

In holometabolous insects, such as butterflies, the adult female must select a plant for oviposition that maximizes fitness of her offspring. Yet she has no ability to obtain direct information on the quality of the food. The larva, in turn, while capable of making food choices in an experimental situation, is often, in practice, without the opportunity to choose. From an evolutionary point of view, there needs to be correspondence of genes expressed in adults and larvae, such that there is a link between adult acceptance of a plant and larval acceptance and fitness on that plant. The adult butterflies tend to be more specific in what they select than are their larvae, and the evidence so far is that different genes are involved in adults and larvae. The separation of the two stages has implications for changes in diet breadth. Since she does not experience any nutritional or noxious effect herself in relation to where she lays eggs, the adult female may readily switch from using several hosts to using a single one, depending on plant abundance (see Section 8.1) She also has a tendency to lay eggs successively on the same plant type (Chapters 4 and 6). If this pattern is repeated over generations in a region, larvae may be selected for fitness parameters on this one plant. If this is accompanied by a reduced fitness on other plants, differential larval fitness may then, in turn, provide a selection pressure on adult behavior. This dual process may hasten change.

Insect lifespans are often short compared with those of host plants they utilize. This would be true for insects with many generations in a year and in insects that feed on perennials. Thus evolutionary change is potentially faster in the herbivore than in the food plant. In a simplistic approach, this might be interpreted as a basis for believing that insects are able to evolve rather quickly to changing needs brought about by changes in the host.

In many groups there is a relationship between diet breadth and number of generations per year. For example, in the moth family Noctuidae, the subfamily Heliothinae is one in which some species are extreme specialists and others are extreme generalists. Specialists, such as those in the genus *Schinia*, have one generation per year, and their phenology matches that of the hosts. Generalists, such as several species of *Heliothis*, have many generations in a year, large geographical ranges, and successive generations must often use different host plants. Whether generalists evolved from specialists, or vice versa, is unknown in this group, but an understanding of the phylogeny would aid our understanding of the selective pressures for different diet breadths. This would allow us to see whether specialists mainly evolve from generalists, or vice versa, or whether evolution occurs commonly in each direction.

Other aspects of insect phenology may be relevant. Several studies have shown that insects diapausing as pupae have a broader diet than those diapausing as eggs or early instar larvae. Among lycaenids, for example, there is a relationship between the stage at which diapause occurs in the insects and the host range (Fig. 8.9). Those that diapause as pupae have about twice as many host genera as those that diapause in the egg or larval stage, or do not diapause at all. A possible interpretation of this difference is that, after a pupal diapause, the adult that emerges is mobile and able to select various hosts according perhaps to

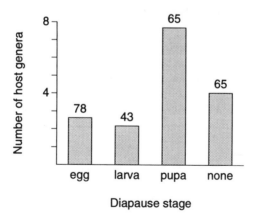

Figure 8.9. Relationship between diapause stage and average diet breadth in lycaenid butterflies. Those diapausing as pupae have broader diets. Numbers above each bar show the number of butterfly species in that category (after Carey, 1992).

quality. On the other hand, after diapause as eggs or first instar larvae, there is usually a very limited choice of potential food. If this interpretation is correct, it indicates that phenology is having a marked effect on diet breadth.

As with many other factors, it is not clear how much the life-history parameters determine diet breadth, or to what extent breadth leads to certain life-history characteristics. Phylogenetic analyses can help to answer this. Fig 8.10 shows two phylogenetic trees for the evolution of a hypothetical group of lepidopterans. In a, the older species, 1,2 and 7 all have a pupal diapause, while the derived species 3 to 6 all have a larval diapause. A change from pupal to larval diapause probably occurred at point *. In such a tree one would conclude that phenology probably influenced diet breadth. On the other hand, in b, no such pattern occurs, and diet breadth might have influenced phenology.

8.4.2 Morphology

Just as at the level of higher taxa, where sucking and chewing orders of insects are limited in the type of food ingested, morphological features of insects effectively limit the range of foods accessible to them. Differences in mouthpart structures provide some of the best examples. Mandibular specializations for specific food types may reduce efficiency of handling a variety of alternative food types. For example, in the moth family Sphingidae, the general pattern of mandible structure is characterized by large blunt, interlocking teeth which appear to be excellent for feeding on soft leaf tissue which is actually torn off and partially shredded during ingestion (Fig. 8.11). Among the giant silk moths, Saturniidae, however, the pattern is of stout toothless mandibles with overlapped

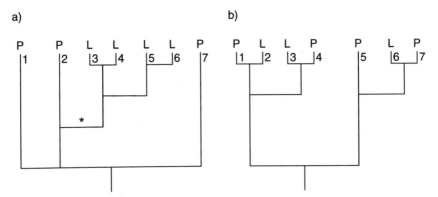

Figure 8.10. Hypothetical phylogenies of 7 species (1-7) of Lepidoptera mapped onto diapause stage. L = larval diapause, P = pupal diapause. **a)** The more ancient lines have a pupal diapause, while the derived ones have a larval diapause. **b)** There is no pattern of diapause stage in relation to age of line or relatedness of the species.

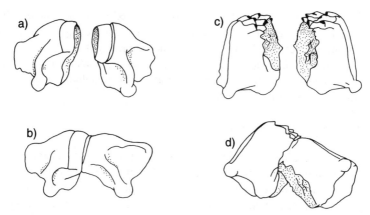

Figure 8.11. Typical mandible shapes of larvae in the lepidopteran families Saturniidae and Sphingidae. **a)**, **b)** Saturniidae. The mandibles are adapted for snipping very tough leaves. **c)**, **d)** Sphingidae. The mandibles are adapted for tearing and shredding soft leaves (after Bernays and Janzen, 1988).

edges, and these are used for snipping off very regular sections of the tough host leaves in the manner of tin snips. Variations in detail occur especially among sphingids, where diet breadth is usually narrow. Certainly, it would be difficult for species adapted to the soft leaf tissues to be able to manage tough leaves, while saturniid mandibles would be quite unsuitable for cutting soft leaves.

In the soapberry bug, *Jadera haematoloma*, populations have evolved different beak lengths to obtain access to seeds in various sizes of fruits. The original population in southern United States had a single species of host, and a mean beak length of the appropriate dimensions. Species of soapberry with large and small fruit sizes have been introduced into different regions of southern United States. Separate populations of the bug have evolved either shorter or longer beaks to obtain access in three species of soapberry. The change has occurred within 50 years as demonstrated by the changes in morphometrics of specimens collected over time and kept in museums. Those with shorter beaks now would have difficulty utilizing the larger fruit. These changes may be a preliminary to differentiation of the soapberry bug into three species with three separate hosts.

Tarsal structures giving specific ability to negotiate a particular kind of leaf surface may be extremely important in small insects. It has been suggested that tarsal morphology of aphids is a basis for specialization on British oaks, since the specialists have tarsi that allow a much better purchase on the leaf surface. In a comprehensive study of morphology in the aphid genus *Uroleucon*, many morphological details can be seen to be important. Leg segment lengths and rostral dimensions vary with host characters, and in one example the rostrum length was shown to be strongly positively correlated with trichome density on the host plant. The pattern was noted in different species complexes within the genus, giving greater weight to its significance (Fig. 8.12).

Although one could enumerate many morphological adaptations, they are likely to be effective in maintaining host affiliation but are unlikely to be causative factors in evolution of diet breadth.

8.4.3 Neurophysiology and behavior

We have seen in Chapter 4 that host-plant selection by phytophagous insects involves not only the acceptance of the host, but also the rejection of nonhosts. Consequently, any move on to a novel host plant necessitates a reduction in this rejection response. Since rejection is a response largely to secondary plant chemicals, a reduction in the behavioral response implies a change in the nervous system. It may occur either peripherally or centrally but is such that the deterrent chemicals in the new host have a reduced effect. In the case of a species evolving into a relative generalist, this can be envisaged as a general reduction in sensitivity since we have seen (Fig. 4.27) that generalist insects are less sensitive to deterrents than their more specific relatives. A change from generalist to specialist on a plant within the host range of the generalist might, in contrast, involve a general

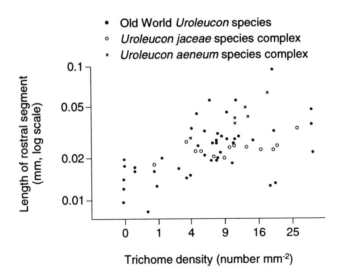

Figure 8.12. Relationship between host-plant trichome density and length of apical rostral segment in aphids of the genus *Uroleucon*. The three different symbols represent three phylogenetically distinct groups of species. A similar relationship between trichome density and rostral segment length occurs in all three (after Moran, 1986).

increase in sensitivity. This is depicted diagrammatically in Fig. 8.13. The generalist, species X, is insensitive and deterred only by compound D. However, a general increase in sensitivity will cause all the compounds to be deterrent except for G. Consequently, the insect will only eat the plant containing this compound; it will reject all the others.

In this second case, the change in the nervous system might only follow a behavioral switch caused by some other factor. However, a shift to a plant not previously in the host range is likely to require a change in the sensory system as a preliminary to establishment on the new host. Unless there is such a change, acceptance of a new plant will be prevented by its deterrent qualities.

How might this happen in practice? The grasshopper, *Bootettix argentatus*, appears to provide an example. This insect feeds only on creosote bush, *Larrea*, despite the fact that all its close relatives in the subfamily Gomphocerinae feed on grasses. We have described, in Section 8.1, the ecological conditions that appear to have brought about this change. Nevertheless, *Larrea* contains a chemical, NDGA, which is deterrent to all the other gomphocerines that have been

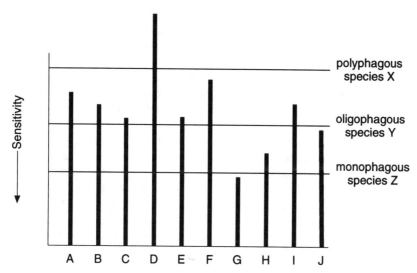

Figure 8.13. Diagrammatic representation of how differences in sensitivity to plant secondary compounds could affect diet breadth. A-J represent different plant secondary compounds. The height of each column represents its stimulating effectiveness as a deterrent at some standard concentration. Because the polyphagous species (X) is relatively insensitive to the compounds, it will feed on plants containing any of them except for D. As the sensitivity of the insect increases (note: downwards) more compounds become deterrent, resulting first in oligophagy (species Y), and finally in monophagy (species Z). For this species only compound G is not deterrent so its host range will be limited to plants containing this compound and no other. All others will be deterrent.

tested (see Fig. 4.26), and we presume this was also true of the ancestors of *B.argentatus*. We picture these ancestors being forced to feed on *Larrea* by the increasing desertification. At first, most individuals died, but a few, with a relatively high level of tolerance, survived, perhaps by selecting parts of the plant with the lowest levels of NDGA. Selection will have favored those individuals with the lowest levels of sensitivity to NDGA, and, in the continuation of this process, sensitivity of the deterrent cell to this compound has disappeared altogether. The phagostimulatory response to NDGA we see today (Fig. 4.26) will have been a later development, involving different chemoreceptor neurons from those mediating deterrence.

Such changes in the patterns of behavior, themselves dependent on chemoreceptor inputs and their integration, do not occur in the absence of other change, but we suggest that they are the first changes necessary in the evolution of an altered host range. Hints of this occur in the literature when ecologists and evolutionary biologists suggest that behavioral changes precede other adaptations in the evolution of specialization. Futuyma, for example, states that specialization is likely to evolve first at a behavioral level, perhaps in response to an abundant resource. He notes that, in terms of optimal foraging theory, specialization when the resource is abundant is an advantage. Jermy suggested that neural elements are the first ones to alter in the development of specialization so that "differentiation of the nervous system leads to specialization of behavior." However, in none of these cases is there any explicit explanation of exactly what may change.

Another element of a neurophysiological approach to understanding diet breadth, concerns what might be described as a limitation in integrative ability. This was first suggested by Levins and MacArthur. They did not have any mechanisms to propose, but they suggested that the extensive incidence of monophagy was due to neural limitations. Many insects can be made to feed on nonhosts, and, with chemoreceptors removed, some will feed and survive on a very wide range of plants, suggesting a basic physiological ability that is limited by the chemoreceptor input and/or the neural processing system. They suggest that this reflects some kind of advantage to limiting the information that has to be processed. In other words, efficiency in host choice behavior is increased by restricting the choice that has to be made. Futuyma has made the same proposal, although the basis for the idea comes from foraging theory rather than from any knowledge of processes.

The significance of the ability of animals to develop a "search image" and/or to pay attention to only the items of interest may be an important factor in the selection of host plants by insects. There is some evidence that it occurs. For example, in chapter 6 the references to butterfly oviposition studies are of particular relevance. In several studies the data indicated that butterflies "learned" to become proficient at selecting the hosts. However, the most detailed study was on lycaenid butterflies with several available host plants, each equally suitable for the larvae. Females tended to land on the same host plant successively, even when the alternative hosts were in the same or higher densities. Furthermore, in a sequence of "oviposition landings," these butterflies were more likely to lay an egg if they landed on the same host in succession, than if they alternated between hosts. This was true whichever host was temporarily favored. An interpretation of this behavior is that the processes of decision-making were simpler when sensory inputs were limited or repetitive, allowing rapid decisions on oviposition to be made in the short periods of landing. In other words, a focussing of the sensory input by selective attention to a subset of the available inputs, would theoretically allow faster decision-making. This would in some sense support the ideas of Levins and McArthur.

8.5 Conclusions

Is there any unifying theme underlying the evolution of specialized diets? After examining the various theories in this chapter it seems unlikely. A variety of factors may be important, and these differ with insect groups. Mobile species that actively choose host plants, especially those that can do so throughout their lives, have different pressures from those that are opportunistic for reasons of immobility and chance dispersal. Furthermore, it is likely that interactions of two or more factors may be much more important than any factor in isolation. An example is the possible combined significance of relative resource availability and predation pressure. Another might be the availability of a large single resource, combined with competitive interactions on the less common plant species. However, in the search for some common factor that may promote specialization the physiological tradeoff idea would be a good candidate with application across taxa, but evidence for this is still poor. Among all these factors, it seems inevitable that many will have played a part in the evolution of narrow diets.

In spite of the potential significance of several of these very different factors, the behavioral/neurological idea of constraints on input and processing of information may be the best candidate for a common factor. Firstly, work on vertebrates indicates quite strongly that decision-making is considerably easier if inputs to the relevant parts of the central nervous system are limited at any one time. The mechanisms involve various ways of channeling information that is perceived, and the terminology includes "paying attention" and "search-image formation." In insects, another method of channelling could involve peripheral sensory interactions or limitations on what is even perceived. Both occur. Secondly, many plant chemicals are deterrent but harmless, and appear to simply be signals. This system of signalling may be an important element in the phylogenetic constraints which seem to exist in terms of ability to recognize or accept potential host plants. Finally, the problems associated with processing information from the very large and complex data set presented by a plant community would be encountered by all herbivorous insects. There is the advantage of parsimony in a theory of diet limitation associated with simplification of decision-making processes.

In coming to a theory of diet breadth based on the properties of the nervous system, we come back once again to the importance of mechanisms. The evolution of behavioral patterns is an evolution of properties of the nervous system. The precise details of how insects perceive the plant world, how they channel and integrate the information, and finally how they behave in response to the information, will provide the details necessary to develop ideas further on how the behavior of host-plant choice in insects may have evolved.

Further reading

Futuyma, D.J. 1991. Evolution of host specificity in herbivorous insects: genetic ecological, and phylogenetic considerations. *In* Price, P.W., Lewinsohn, T.M., Fernandes, G.W. and Benson, W.W. (eds.) *Plant-Animal Interactions: Evolutionary Ecology in Tropical and Temperate Regions*. Wiley, New York, pp. 431–454.

Futuyma D.J. and Moreno, G. 1988. The evolution of ecological specialization. A. Rev. Ecol. Syst. 19: 203–233.

Futuyma, D.J. and Peterson, S.C. 1985. Genetic variation in the use of resources by insects. A.Rev.Entomol. 30: 217–238.

Jaenike, J. 1990. Host specialization in phytophagous insects. A.Rev.Ecol.Syst. 21: 243–274.

*McClure, M.S. 1983. Competition between herbivores and increased resource heterogeneity. *In* Denno, R.F. and McClure, M.S. (eds.) *Variable Plants and Animals in Natural and Managed Systems*. Academic Press, New York, pp. 125–154.

Strong, D.R, Lawton, J.H and Southwood, R. 1984. *Insects on Plants*. Harvard University Press, Cambridge.

References (* indicates review)

Scale

Carey, D.B. 1992. Factors determining host plant range in two lycaenid butterflies. Ph.D. thesis, University of Arizona.

Courtney, S.P and Chew, F.S. 1987. Coexistence and host use by a large community of pierid butterflies: habitat is the templet. Oecologia 71: 210–220.

*Gaston, K.J.and Lawton, J.H. 1988. Patterns in body size, population dynamics, and regional distribution of bracken herbivores. Am.Nat. 132: 662–680.

*Gaston, K.J. and Reavey, D. 1989. Patterns in the life histories and feeding strategies of British macrolepidoptera. J.Linn.Soc. 37: 367–381.

*Kogan, M. and Turnipseed, S.G. 1987. Ecology and management of soybean arthropods. Ann.Rev.Entomol.32: 507–538

*Mattson, W.B. 1980. Herbivory in relation to plant nitrogen content. A.Rev.Ecol.Syst. 11: 119–161.

Mitter, C., Futuyma, D.J., Schneider, J.C. and Hare, J.D. 1979. Genetic variation and host plant relations in a parthenogenetic moth. Evolution 33: 777–790.

Opler, P. 1974. Oaks as evolutionary islands for leaf-mining insects. Am.Scient. 62: 67–73.

Otte, D. and Joern, A. 1977. On feeding patterns in desert grasshoppers and the evolution of specialized diets. Proc.Acad.Nat.Sci.Phil. 128: 89–126.

*Rathke, B. and Lacey, E.P. 1985. Phenological patterns of terrestrial plants. A. Rev. Ecol. Syst. 16: 179–214.

*Southwood, T.R.E. 1972. The insect/plant relationship—an evolutionary perspective. Symp.R.Entomol.Soc.Lond. 6: 3–30.

Wasserman, S.S. and Mitter, C. 1978. The relationship of body size to breadth of diet in some Lepidoptera. Ecol.Entomol. 3: 155–160.

Wiklund, C. 1974. The concept of oligophagy and the natural habitat and host plants of *Papilio machaon* in Fennoscandia. Entomol.Scand. 5: 151–160.

Interactions among organisms

Bernays, E.A. 1989. Host range of phytophagous insects: the potential role of generalist predators. Evolutionary Ecology 3: 299–311.

Bernays, E.A. and Graham, M. 1988. On the evolution of host specificity in phytophagous arthropods. Ecology 69: 886–892.

Brower, L.P. 1958. Bird predation and foodplant specificity in closely related procryptic insects. Am.Nat. 864: 183–187.

*Colwell, R.K. 1985. Community biology and sexual selection: lessons from hummingbird flower mites. *In* Diamond, J. and Case, T.J. (eds.) *Community Ecology*. Harper & Row, New York, pp. 406–424.

Damman, H. 1987. Leaf quality and enemy avoidance by larvae of a pyralid moth. Ecology 68: 87–97.

*Damman, H. 1992. Patterns of interaction among herbivore species. *In* Stamp, N.E. and Casey, T.M. (eds.) *Caterpillars*. Chapman and Hall, New York, pp. 132–169.

Denno, R.F., Larsson, S. and Olmstead, K.L. 1990. Role of enemy-free space and plant quality in host-plant selection by willow beetles. Ecology 71: 124–137.

Eastop, V.F. 1973. Deductions from the present day host plants of aphids and related insects. Symp.R.Entomol.Soc.Lond.6: 157–178.

*Fritz, R.S. 1992. Community structure and species interactions of phytophagous insects on resistant and susceptible host plants. *In* Fritz, R.S. and Simms, E.L. (eds.) *Plant Resistance to Herbivores*. University of Chicago Press, Chicago, pp. 240–276.

Heads, P.A. and Lawton, J.G. 1985. Bracken, ants and extrafloral nectaries.II. The effect of ants on the insect herbivores of bracken. J.Anim.Ecol. 53: 1015–1031.

Jaenike, J. 1986. Parasite pressure and the evolution of amanitin tolerance in *Drosophila*. Evolution 39: 1295–1302.

McClure, M.S. 1980. Competition between exotic species: scale insects on hemlock. Ecology 61: 1391–1401.

Moran, N.A. 1986. Benefits of host plant specificity in *Uroleucon* (Homoptera: Aphididae). Ecology 67: 108–115.

Stamp, N.E. and Bowers, M.D. 1991. Indirect effect on survivorship of caterpillars due to presence of invertebrate predators. Oecologia 88: 325–330.

*Ward, S. 1991. Reproduction and host selection by aphids: the importance of "rendezvous" hosts. *In* Bailey, W.J. and Ridsill-Smith, J. (eds.) *Reproductive Behavior of Insects*. Chapman and Hall, New York, pp. 202–226.

Woodman, R.L. and Price, P.W. 1992. Differential larval predation by ants can influence willow sawfly community structure. Ecology 73: 1028–1037.

Zwolfer, H. 1979. Strategies and counterstrategies in insect population systems competing for space and food in flower heads and plant galls. Fortschr.Zool. 25: 331–353.

Plant Chemistry

*Berenbaum, M. 1990. Evolution of specialization in insect-umbellifer associations. A.Rev.Entomol. 35: 319–343.

Bernays, E.A. 1990. Plant secondary compounds deterrent but not toxic to the grass specialist *Locusta migratoria*: implications for the evolution of graminivory. Entomologia Exp.Appl. 54: 53–56.

*Bernays, E.A. and Chapman, R.F. 1987. The evolution of deterrent responses in plant-feeding insects. *In* Chapman, R.F., Bernays, E.A. and Stoffolano, J.G. (eds.) *Perspectives in Chemoreception and Behavior*. Springer Verlag, New York, pp. 159–173.

Fry, J.D. 1990. Trade-offs in fitness on different hosts: evidence from a selection experiment with a phytophagous mite. Amer.Nat. 136: 569–580.

Futuyma, D. and McCafferty, S.J. 1990. Phylogeny and the evolution of host plant associations in the leaf beetle genus *Ophraella* (Coleoptera: Chrysomelidae). Evolution 44: 1885–1913.

Futuyma, D.J and Wasserman, S.S. 1981. Food plant specialization and feeding efficiency in the tent caterpillars *Malacosoma disstria* and *Malacosoma americanum*. Entomologia Exp.Appl. 30: 106–110.

Gould, F. 1979. Rapid host range evolution in a population of a phytophagous mite *Tetranychus urticae* Koch. Evolution 33: 791–802.

*Jaenike, J. 1990. Host specialization in phytophagous insects. A. Rev. Ecol. Syst. 21: 243–274.

Karowe, D.N. 1989. Facultative monophagy as a consequence of prior feeding experience: behavioral and physiological specialization in *Colias philodice* larvae. Oecologia 78: 106–111.

*Menken, S.B.J., Herrebout, W.M. and Wiebes, J.T. 1992. Small ermine moths (*Yypono-meuta*): their host relations and evolution. A.Rev.Entomol. 37: 41–66.

*Mitter, C. and Farrell, B. 1991. Macroevolutionary aspects of insect-plant interactions. *In* Bernays, E.A. (ed.) *Insect Plant Interactions* Vol III. CRC Press, Boca Raton, pp. 35–78.

Rausher, M.D. 1984. Tradeoffs in performance on difference hosts: evidence from within and between-site variation in the beetle *Deloyala guttata*. Evolution 38: 582–595.

*Zwolfer, H. 1982. Patterns and driving forces in the evolution of plant-insect systems. *In* Visser, J.H. and Minks, A.K. (eds.) *Insect-Plant Relationships*. Pudoc, Wageningen, pp. 287–296.

Insect characteristics

*Bernays, E.A. 1991. Evolution of insect morphology in relation to plants. *In* Chaloner, W.G., Harper, J.L. and Lawton, J.H. (eds.) *The Evolutionary Interaction of Animals and Plants*. Royal Society, London, pp. 81–88.

*Bernays, E.A. and Chapman, R.F. 1987. The evolution of deterrent responses in plant-feeding insects. In Chapman, R.F., Bernays, E.A. and Stoffolano, J.G. (eds.) *Perspectives in Chemoreception and Behavior*. Springer Verlag, New York, pp. 159–173.

Bernays, E.A. and Janzen, D. 1988. Saturniid and sphingid caterpillars: two ways to eat leaves. Ecology 69: 1153–1160.

Carey, D.B. 1992. Factors determining host plant range in two lycaenid butterflies. Ph.D. thesis, University of Arizona.

Carroll, S. P. and Boyd, C. 1992. Host race radiation in the soapberry bug: natural history with the history. Evolution 46: 1053–1069.

*Futuyma, D.J. 1983. Selective factors in the evolution of host choice by phytophagous insects. *In* Ahmad, S. (ed.) *Herbivorous Insects; Host Seeking Behavior and Mechanisms*. Academic Press, New York, pp. 227–245.

*Jermy, T., Labos, E. and Molnar, I. 1990. Stenophagy of phytophagous insects—a result of constraints on the evolution of the nervous system. *In* Maynard Smith, J. and Vida, G. (eds.) *Organizational Constraints on the Dynamics of Evolution*. Manchester University Press, Manchester, pp. 157–166.

Kennedy, C.E.J. 1987. Attachment may be a basis for specialization in oak aphids. Ecol.Entomol. 11: 291–300.

*Levins, R. and MacArthur, R. 1969. An hypothesis to explain the incidence of monophagy. Ecology 50: 910–911.

Moran, N.A. 1986. Morphological adaptation to host plants in *Uroleucon* (Homoptera: Aphididae). Evolution 40: 1044–1050.

*Thompson, J.N. and Pellmyr, O. 1991. Evolution of oviposition behavior and host preference in Lepidoptera. A.Rev.Entomol. 36: 65–89.

*Zwolfer, H. 1987. Species richness, species packing, and evolution in insect-plant systems. Ecological Studies 61: 301–319.

Glossary

Abaxial (surface): directed away from axis; e.g., lower surface of normal leaves

Acceptability: acceptance of item for feeding or oviposition

Acceptor site: site on sensory neuron where chemical that induces a receptor potential interacts with the membrane

Adaptation: reduced activity of sensory neuron over time to a sustained stimulus

Adaxial (surface): directed towards axis; e.g. the upper surface of normal leaves

Airspeed: speed of a moving object, e.g., a flying insect, through the air

Allele: any of the different forms of a gene occupying the same locus

Allelochemical: a chemical produced by an organism that produces an effect in a different species. In phytophagous insects, this often means an effect on behavior, growth or development. Plant secondary metabolites are often called allelochemicals

Allelopathy: noxious effect of plant secondary metabolite on neighboring plants of different species

Allozymes: the various forms of an enzyme which have the same activity but which differ slightly in amino acid sequence, produced by different alleles at a single locus.

Anemotaxis: orientation to wind

Angiospermae: one of the two great divisions of seed plants (Spermatophyta); distinguished from Gymnospermae by the presence of true flowers with ovaries

Antifeedant: chemical that inhibits feeding = deterrent

Appetitive behavior: active "goal-seeking" behavior; not a preferred term

Arousal: a change in activity in part of the central nervous system leading to an increased responsiveness to various stimuli. Usually based on observations of behavior

Arrestant: chemical (usually) that inhibits further locomotion

Attractant: chemical causing orientation and movement towards source

Associative learning: learned association between two sensory inputs, one of which is physiologically or behaviorally significant, while the other normally has no specific contextual significance

Aversion learning: a learned association between a sensory cue which is otherwise meaningless, and a noxious effect

Ballooning: aerial rising and dispersal with the help of silk threads, common in newly hatched caterpillars and spiders

Biomass: quantitative estimate of the total mass of organisms of specified unit in a specified area and time.

Biotype: an infraspecific category distinguishable by biochemistry or behavior, and sometimes equivalent to "race"

C3 plant: plant with a photosynthetic pathway of the normal Calvin type, via ribulose bisphosphate and glycollate, with a proportion of the CO_2 being lost

C4 plant: plant with an accessory photosynthetic pathway of the Hatch/Slack type, in which CO_2 is returned to the Calvin cycle. This occurs in plants adapted to hot, dry conditions; associated anatomy involves thickened cells in the sheath surrounding the vascular tissue. These cells are supposedly difficult for insects to penetrate

Cellulose: carbohydrate consisting of long chain molecules comprising anhydrous glucose residues as basic units; a principal constituent of cell walls

Chemoreception: detection of chemicals by smell or taste

Chemotaxis: orientation to a chemical

Chemotaxonomy: taxonomic grouping based on chemical characters

Compensatory feeding: an increase in feeding to make up for low levels of a particular nutrient

Conditioning: learning an association

Congeneric: species in the same genus

Conspecific: same species

Cryptic: similar to background and consequently making the object difficult to discern

Cultivar: a variety of a plant species produced and maintained by cultivation

Cuticle: layer of cutin in plants, a fatty substance which is almost impermeable to water; present on outer walls of epidermal cells, and usually with a separate wax layer on the outside

Deme: a local interbreeding group

Desensitization: = habituation

Desmotubule: see plamodesmata

Deterrent: chemical which inhibits feeding or oviposition when present where feeding or oviposition would otherwise occur.

Detoxification enzyme: enzyme, often in the gut tissue, that makes toxic substances less noxious

Diapause: a delay in insect development that is not immediately referable to environmental conditions and that enables the insect to survive predictable periods of adverse conditions such as winter or a dry season

Diet breadth: number of host species fed upon

Disadaptation: return of sensitivity after adaptation of a sensory receptor

Electrophoresis: a laboratory technique used for separating proteins and often employed for separating allozymes, thus demonstrating genetic variability which is not measurable by other means. Its widespread use first demonstrated a previously unknown degree of variation in wild populations

Endophyte: fungal organism living within a green plant often with little or no measurable symptomatic effects on the host

Epiphyte: plants, including fungi, mosses and algae, growing on the external surface of vascular plants

Euryphagous: polyphagous

Excitation: increased activity apparently denoting a heightened level of arousal

Fatigue (in chemoreceptor): short term inability to respond to stimuli due to non-renewal of part of a chemical pathway

Feeding inhibitor: antifeedant or deterrent

Fitness: the relative competitive ability of a given genotype conferred by adaptive characters, usually quantified as the number of surviving progeny relative to other genotypes. It is a measure of the contribution of a given genotye to the subsequent generation relative to that of other genotypes

Folivore: animal feeding on leaves (foliage)

Forb: herbaceous broad-leaved plant (nongrass)

Forbivorous: feeding on forbs

Generalist: polyphagous species

Genetic drift: random changes in gene frequencies in isolated populations not due to selection pressure

Genotype: the genetic constitution of an individual (cf. phenotype)

Glucoside: compound with glucose moiety

Glycoside: compound with sugar moiety; many different sugars are involved including glucose, rhamnose, xylose

Graminivorous: herbivores that feed only on grasses

Granivorous: animals feeding on small grains; cereal and legumes are both included.

Gravid: term applied to a female insect containing fully developed eggs ready to be laid

Groundspeed: speed of movement of a flying object over the ground

Guttation: secretion of drops of water by plants through hydathodes

Habituation: waning of a response to a stimulus

Headspace: the volume of air around a plant which can be removed and analyzed to determine the volatiles produced

Hemimetabolous: insect development involving a partial metamorphosis as in grasshoppers and Hemiptera

Heritability (h^2): the part of phenotypic variability that is genetically based; usually expressed as the ratio of genetic variance to phenotypic variance

Heterotherm: animal whose body temperature depends partly on the environmental temperature, but can sometimes be regulated physiologically

Heterozygous: having two different alleles at a given locus of a chromosome pair

Holometabolous: insect development involving a complete metamorphosis with a pupal stage interpolated between larva and adult, as in Lepidoptera, Coleoptera, Diptera and Hymenoptera

Homeotherm: animal which regulates its body temperature within narrow limits independent of environmental temperature

Honeydew: waste material produced by aphids and other phloem-feeding insects, and usually containing high levels of sugars

Hormone: organic substance produced in small quantity in one part of an organism, and transported to other parts where it exerts an effect on physiology or behavior

Host range: relative number of host-plant species utilized by a herbivore

Ideoblast: a plant cell which markedly differs in form, size or contents from other cells in the same tissue

Imprinting: a persisting effect of a particular experience at some specific stage that alters behavior. Usually at an early stage of development

Induction: usually meaning an increased acceptance of a plant species that the insect has recently eaten, relative to other plants which have not been eaten

Insolation: solar radiation

Instar: stage of an arthropod between two molts

Isofemale strain: a strain derived from a single fertilized founder female

Kairomone: chemical released by an organism and adaptively useful to other organisms that perceive it.

Kinesis: change in activity level

Latex: a fluid, generally milky, contained in laticifers, and consisting of a variety of organic and inorganic substances which together congeal on exposure to air

Laticifer: cell or series of cells with characteristic latex fluid content; usually tubular in shape, may be branched or unbranched

Lignin: an organic complex of high carbon-content substances, especially phenolics, present in cell walls

Metamorphosis: a marked structural transformation during development, often representing a change from larval stage to adult. In insects, used to refer to the change from the final larval stage to the adult

Mixed function oxidase: enzyme which has among its effects an ability to alter (detoxify) xenobiotics. More commonly called P450 enzymes because they absorb light of 450 nm wavelength

Monophagous: feeding on a single species or genus of plants

Monophyletic group: a group of taxa descended from a single ancestral taxon

Motivation: term used to imply the physiological state of an animal that predisposes it to carry out certain behaviors, e.g., feeding

Nectary: multicellular, glandular structure in plants capable of secreting sugary solution. Floral nectaries occur in flowers; extrafloral nectaries occur in other plant organs

Neophilia: predilection for novelty

Neophobia: distaste of novelty

Olfaction: smell; detection of volatile chemicals

Oligolectic: limited host range; used for pollinators, especially solitary bees

Oligophagous: eating a restricted number of plant species usually from a family or subfamily of plants

Ontogeny: development

Oviposition: egg-laying

Oviposition stimulant: chemical or other stimulus eliciting oviposition

Palatability: acceptability as food

Parasitoid: an organism with a mode of life intermediate between parasitism and predation. Usually a fly (Diptera) or wasp (Hymenoptera) in which the larva feeds within the body of the live host, eventually killing it

Parthenogenetic: production of offspring in the absence of fertilization

Pectic compounds: polymers of galacturonic acid and its derivatives; main constituent of lamella between plant cells

Phagostimulant: chemical eliciting feeding

Phenology: temporal aspects of the life history

Phenotype: the observable structural and functional properties of an organism (cf. genotype)

Pheromone: chemical produced by an animal which influences the behavior or physiology of conspecifics

Phloem: the principal food-conducting tissue of the vascular plant, nutrients produced from photosynthesis are here transported, usually under positive pressure away from leaves; particularly rich in sugars

Phylloplane: the leaf surface

Phylogenetic tree: a branching diagram in the form of a tree representing inferred lines of descent

Phytophagous: feeding on plants

Plasmodesmata: thin cytoplasmic strands passing through pores in the plant cell wall and usually connecting the protoplasm of adjacent cells

Poikilotherm: animal with variable temperature dependent on environmental temperature and radiation

Polygenic traits: those influenced by allele differences at several or many loci

Polyphagous: eating a wide spectrum of plants across several or many families

Polytrophy: feeding on many items

Population genetics: evaluates allele frequency change without regard to phenotypes. It predicts allele frequencies under different environmental circumstances

Preference: selection of item from a choice of items

Quantitative genetics: evaluates phenotypic change without regard to identity of alleles. It predicts and measures response to selection of polygenic traits

Raphide: needle-shaped crystal of calcium oxalate; usually one of a number of crystals arranged parallel to one another in a plant cell

Receptor: sensory neuron

Repellant: causes movement away from source

Resistance: ability to withstand attack

Rhizosphere: soil immediately surrounding plant roots and influenced by them

Sclerenchyma: a mechanical or supporting tissue in plants, made up of cells with lignified walls

Sclerophyll: see sclerenchyma

Search image: a model for visual comparison, and presumed to account for increased efficiency in finding items

Secondary metabolite: chemical produced by a plant that is not essential for primary metabolic processes

Self-selection: a term sometimes used to denote the selection of different foods that will result in a balance of nutrients ingested

Sensitization: increased responsiveness without learning

Sequester: active accumulation of material, usually a plant secondary chemical, by a herbivore, and often used as a means of protection from higher trophic level organisms

Sign stimulus: a stimulus, or one of a group of stimuli, by which an animal distinguishes important objects. Often used for chemicals that indicate a specific host

Silica body: opaline cell inclusion in plant epidermal cells; shapes of silica bodies may be characteristic for a family or lower taxon; common in grasses

Specialization: evolutionary adaptation to a particular mode of life or habitat

Specificity: degree of restriction of host-plant range

Starch: an insoluble carbohydrate acting as one of the commonest storage products of plants, composed of anhydrous glucose residues

Stenophagous: utilizing a limited variety of foods

Stochastic: due to chance alone

Stoma (plural stomata): pore in the epidermis of plants encircled by two guard cells which regulate its size

Symbiosis: the living together of two organisms; sometimes restricted to those that are mutually beneficial

Sympatric: populations, species or other taxa, occurring together in the same geographic area

Sympatric speciation: the differentiation and attainment of reproductive isolation of populations that are not geographically separated, and which overlap in their distributions

Synapse: point of communication between two neurons, usually requiring the release of a particular neurotransmitter, that conveys the signal from one neuron to the next

Synergism: action of two or more chemicals such that the total effect is greater than the sum of the individual effects

Synomone: chemical released by one organism that benefits the producing organism and a receiving organism of a different species

Taxis: an orientated response towards or away from the source of stimulus

Testa: seed coat

Toxicity: the virulence of a toxic substance

Toxin: a biogenic poison, usually proteinaceous

Trichome: epidermal appendage in plants; includes hairs, scales and papillae; may be glandular or nonglandular

Tritrophic: three trophic levels; as in plant-herbivore-predator

Umbel: inflorescence in which branches ending in flowers all arise at the same point on the stem, as in carrot, celery, parsley

Vacuole: a membrane-bound cavity in the cytoplasm; large in plants and often containing secondary metabolites

Variety: a rank in the hierarchy of plant classification below subspecies

Vascular: in plants, referring to the phloem and xylem

Vascular bundle: an organized strand of phloem and xylem
Vein: a vascular bundle or group of parallel bundles in a leaf, sepal, stem

Windspeed: speed of air movement relative to the ground
Woody plant: plant with high levels of lignin providing structural support for growth to shrubs and trees

Xenobiotic: compound that is foreign to the organism, usually assumed to be noxious
Xylophagous: wood feeding
Xylem: plant vascular tissue in which water and minerals are transported from root to shoot; usually under negative pressure

Taxonomic Index*

*References to illustrations or tables are italicized unless there are two or more successive pages

Subject Index*

*References to illustrations and tables are italicized unless there are two or more successive pages